Journeyman Electrician's Review

FIFTH EDITION

Journeyman Electrician's Review

FIFTH EDITION

Richard E. Loyd

THOMSON
DELMAR LEARNING ™

Australia   Canada   Mexico   Singapore   Spain   United Kingdom   United States

THOMSON

DELMAR LEARNING

Journeyman Electrician's Review, 5th Edition
Richard E. Loyd

Vice President, Technology and Trades SBU:
Alar Elken

Editorial Director:
Sandy Clark

Acquisitions Editor:
Patrick Kane

Development Editor:
Jennifer A. Thompson

Marketing Director:
David Garza

Channel Manager:
William Lawrensen

Marketing Coordinator:
Casey Bruno

Production Director:
Mary Ellen Black

Production Manager:
Larry Main

Production Editor:
Ruth Fisher

Library of Congress Cataloging-in-Publication Data:

Loyd, Richard E.
    Journeyman electrician's review / Richard E. Loyd. — 5th ed.
        p. cm.
    Includes index.
    ISBN 1-4018-7949-7
    1. Electric engineering—Examinations, questions, etc. 2. Electricians—Licenses—United States. 3. Electric engineering—Problems, exercises, etc. I. Title.
TK169.L69 2005
621.319'24—dc22
                            2004059884

NOTICE TO THE READER

Publisher does not warrant or guarantee any of the products described herein or perform any independent analysis in connection with any of the product information contained herein. Publisher does not assume, and expressly disclaims, any obligation to obtain and include information other than that provided to it by the manufacturer.

The reader is expressly warned to consider and adopt all safety precautions that might be indicated by the activities herein and to avoid all potential hazards. By following the instructions contained herein, the reader willingly assumes all risks in connection with such instructions.

The publisher makes no representation or warranties of any kind, including but not limited to, the warranties of fitness for particular purpose or merchantability, nor are any such representations implied with respect to the material set forth herein, and the publisher takes no responsibility with respect to such material. The publisher shall not be liable for any special, consequential, or exemplary damages resulting, in whole or part, from the readers' use of, or reliance upon, this material.

I dedicate this book to my wife and business partner, Nancy. She is my inspiration and number one fan. Her assistance is invaluable. I would also like to thank all my wonderful circuit rider friends. Their expertise in the *National Electrical Code®* and the electrical industry continue to provide me with the latest information on our great industry. Finally, to Earl Featherston, my apprenticeship instructor in Pocatello, Idaho, IBEW local union 449, over 40 years ago. Earl put up with us and taught us from the first class to the last class that we must be able to understand and apply the rules of the *National Electrical Code®*.

Table of Contents

CHAPTER FOURTEEN

Foreword

ELECTRIC PER SE

Electricity! In ancient times it was believed to be an act of the gods. Not even Greco-Roman civilizations could understand this thing we know as electricity. It was not until about A.D. 1600 that any scientific theory was recorded; after 17 years of research, William Gilbert wrote a book on the subject titled *De Magnete*. It was almost another 150 years before major gains were made in understanding electricity so that people might some day be able to harness it. At that time, Benjamin Franklin, often referred to as the grandfather of electricity or electrical science, and his close friend, Joseph Priestley, a great historian, began to gather the works of the many worldwide scientists who had been working independently. (One has to wonder where this wondrous industry would be today if the great minds of yesteryear had had the benefit of the great communication networks we have today.) Benjamin Franklin traveled to Europe to gather this information, and with that trip our industry began to surge forward for the first time. Franklin's famous kite experiment allegedly took place about 1752. Coupling the results of that with other scientific data gathered, Franklin decided

he could sell installations of lightning protection to every building owner in the city of Philadelphia. He was very successful in doing just that! (We credit Franklin for the terms "conductor" and "nonconductor," replacing "electric per se" and "non-electric."

With the accumulation of the knowledge gained from these many scientists and inventors of the past, many other great minds emerged and lent their names to even more discoveries. Their names are familiar as other terms related to electricity. Alessandro Volta, James Watt, André Marie Ampère, George Ohm, and Heinrich Hertz are names of just a few of the great minds that enabled our industry to come together.

The next major breakthrough came only a little more than 100 years ago. Thomas Edison, the holder of hundreds of patents promoting direct current (dc) voltage, and Nikola Tesla, inventor of the 3-phase motor and the promoter of alternating current (ac) voltage, began an inimical (bitter) competition in the late 1800s. Edison was methodically making inroads with direct current all along the Eastern Seaboard, and

Electricity is here! *(Reprinted with permission by Underwriters Laboratories, Inc.®)*

The 1893 Great Columbian Exposition at the Chicago World's Fair. The Palace of Electricity astonished crowds; it also astonished electricians by repeatedly setting fire to itself. *(Reprinted with permission by Underwriters Laboratories, Inc.®)*

Tesla had exhausted his finances for his experimental project with alternating current in Colorado. Then the city of Chicago requested bids to electrically light the great Columbian Exposition of 1893. Lighting the world's fair in Chicago, celebrating 300 years since Columbus discovered America, required building generators and nearly a quarter million lights. Edison teamed up with General Electric and submitted a bid based on the Edison lightbulb powered by direct current. George Westinghouse, already successful in the electrified railroad, the elevator, and about 400 other inventions, joined with Nikola Tesla and submitted the lowest bid—about one third of the Edison-General Electric bid. Westinghouse and Tesla were awarded the contract. They based their bid on the Westinghouse Stopper Light, powered by alternating current. The installation was made by Guarantee Electric of Saint Louis, Missouri, which is now one of the oldest and largest electrical contractors in the world. It is said that thousands came out to see this great lighting display. Almost as spectacular as the lighting were the arcing, sparking, and fires started by the display. Accidents occurred daily! It is said that Edison was embellishing the dangers of alternating current by electrocuting dogs and promoted direct current as being much safer. However, the public knew better, because plenty of fires and accidents were also occurring with direct current.

About this time three very important events took place almost simultaneously. The Chicago Board of Fire Underwriters hired William Henry Merrill, an electrician from Boston, as an electrical inspector for the exposition. Merrill saw the need for safety inspections and started Underwriters Laboratories in 1894. In 1895 the first nationally recommended code was published by the National Board of Fire Underwriters (now the American Insurance Association). The 1895 *National Electrical Code®* (*NEC®*)* was drafted through the combined efforts of the architectural, electrical, insurance, and other allied interests based on a document that resulted from actions taken in 1892 by Underwriters National Electrical Association, which met and consolidated various codes that were already in existence. The first *National Electrical Code®* (*NEC®*) was presented to the National Conference on Electrical Rules, which was composed of delegates from various national associations who voted to unanimously recommend it to their respective associations for adoption or approval. A group from the city of Buffalo, New York, visited the world's fair and, based on that visit, hired George Westinghouse and Nikola Tesla to build ac hydro-generators to be powered by Niagara Falls. Alternating current (ac) had won out over direct current (dc) as this country's major power source. The rest is history!

*National Electrical Code® and NEC® are registered trademarks of the National Fire Protection Association, Inc., Quincy, MA 02269.

Preface

The primary purpose of *Journeyman Electrician's Review* is to provide electrical students with a concise, easily understandable study guide to the 2005 edition of the *National Electrical Code®* and the application of electrical calculations. This text is written recognizing that the electrical field has many facets, and the user may have diverse interests and varying levels of experience. However, the interests of the relatively experienced electrician preparing to enter the electrical industry as an entry-level journeyman electrician, the electrician-in-training, and especially the advanced electrical student apprentice have been carefully considered in the text's preparation. *Journeyman Electrician's Review* provides a ready source of the basic information on the *NEC®*. This text may be used for reference or as study material for students and electricians who are preparing for an examination for licensing. It is especially designed as an aid to those studying for the nationally recognized examinations, such as Experior and the Southern Building Code Congress International (SBCCI). It will also help anyone preparing for any electrical examination, and it will provide a quick, easily understood study guide for those needing to update themselves on the *National Electrical Code®* and basic electrical mathematical formulas and calculations. The text is brief and concise for easy application for classroom or home study.

> Each lesson is designed purposely to require the student to apply the entire *NEC®* text, and not specific chapters or articles. It has been found that when studying for a timed, open-book examination, the student must gain proficiency in the Table of Contents, the Index, and the ability to move quickly from cover to cover to find the correct answer to each question in a timely fashion. Check with the testing organization to see whether you are permitted to add index tabs or highlight important sections. If permitted, this can speed up your search for the information.

METRICS (SI) AND THE *NEC®*

The United States is the last major country in the world not using the metric system as the primary system of measurement. We have been very comfortable using English (U.S. Customary) values. This is changing.

STUDY RECOMMENDATIONS

- Read Chapter One carefully.
- Take the Sample Examination. The most important point is that you treat the test the same way you would the "real test." Allow yourself three hours, and grade yourself honestly. This examination will point out to you the areas where you are weak—for example, speed, grounding, or services.
- Study Chapter Two through Chapter Thirteen and take the Question Review following each chapter.
- Take the 50-question Practice Examination in Chapter Fourteen. Allow yourself two hours for the examination.
- Evaluate the areas in which you are still weak.
- Take the Practice Examinations in Chapter Fourteen until you feel you have overcome your weak areas.
- Take the Final Examination in Chapter Fourteen. Allow yourself three hours. If you have studied, you should be prepared. Good luck!

Manufacturers are now showing both inch-pound and metric dimensions in their catalogs. Plans and specifications for governmental new construction and renovation projects begun after January 1, 1994, have been done using the metric system. You may not feel comfortable with metrics, but metrics are here to stay. You might just as well get familiar with the metric system.

The *NEC®* and other *NFPA* Standards are becoming international standards. All measurements in the 2005 *NEC®* are shown with metrics first, followed by the inch-pound value in parentheses—for example, 600 mm (24 in.). Ease in understanding is of utmost importance in *Journeyman Electrician's Review*. Therefore, inch-pound values are shown first, followed by metric values in parentheses—for example, 24 inches (600 mm). A *soft metric conversion* is when the dimensions of a product already designed and manufactured to the inch-pound system are converted to metric dimensions. The product does not change in size. A *hard metric measurement* is when a product has been designed to SI metric dimensions. No conversion from inch-pound measurement units is involved. A *hard conversion* is when an existing product is redesigned into a new size.

In the 2005 edition of the *NEC®*, existing inch-pound dimensions did not change. Metric conversions were made, then rounded off. Where rounding off would create a safety hazard, the metric conversions are mathematically identical. For example, if a dimension is required to be 6 feet, it is shown in the *NEC®* as 1.8 m (6 ft). Note that the 6 ft remains the same, and the metric value of 1.83 m has been rounded off to 1.8 m. *Journeyman Electrician's Review* reflects these rounded off changes, except that the inch-pound measurement is shown first—that is, 6 feet (1.8 m).

TRADE SIZES

A unique situation exists. Strange as it may seem, what electricians have been referring to for years has not been correct! Raceway sizes have always been an approximation. For example, there has never been a ½ inch raceway! Measurements taken from the *NEC®* for a few types of raceways show the following:

Trade Size	Inside Diameter
½ Electrical Metallic Tubing	0.622 inch
½ Electrical Nonmetallic Tubing	0.560 inch
½ Flexible Metal Conduit	0.635 inch
½ Rigid Metal Conduit	0.632 inch
½ Intermediate Metal Conduit	0.660 inch

You can readily see that the cross-sectional areas, critical when determining conductor fill, are different. It makes sense to refer to conduit, raceway, and tubing sizes as "**trade sizes.**" *NEC®* 90.9(C)(1) states that *where the actual measured size of a product is not the same as the nominal size, trade size designators shall be used rather than dimensions. Trade practices shall be followed in all cases.** *Journeyman Electrician's Review* uses the term "**trade size**" when referring to conduits, raceways, and tubing. For example, instead of referring to a ½ inch EMT, it is referred to as trade size ½ EMT.

The *NEC®* also uses the term **metric designator**. A ½ inch EMT is shown as metric designator 16 (½). A 1 inch EMT is shown as metric designator 27 (1).

* Reprinted with permission from NFPA 70-2005.

The numbers 16 and 27 are the metric designator values. The (½) and (1) are the trade sizes. The metric designator is the raceways' inside diameter, in rounded off millimeters (mm). Here are some of the more common sizes of conduit, raceways, and tubing. A complete table is found in the *NEC®*, Table 300.1(C). Because of possible confusion, this text uses only the term "trade size" when referring to conduit and raceway sizes.

METRIC DESIGNATOR AND TRADE SIZE	
Metric Designator	Trade Size
12	⅜
16	½
21	¾
27	1
35	1¼
41	1½
53	2
63	2½
78	3

Conduit knockouts in boxes do not measure up to what we call them. Here are some examples.

Trade Size Knockout	Actual Measurement
½	⅞ inch
¾	1³⁄₃₂ inches
1	1⅜ inches

Outlet boxes and device boxes use their nominal measurement as their trade size. For example, a 4 inch × 4 inch × 1½ inch does not have an internal cubic-inch volume of 4 × 4 × 1½ = 24 cubic inches. Table 370.16(A) shows this size box as having a volume of 21 cubic inches. This table shows trade sizes in two columns in both millimeters and inches.

In this text, a square outlet box is referred to as 4 × 4 × 1½ inch square, 4 inch × 4 inch × 1½ inch square, or trade size 4 × 4 × 1½. Similarly, a single-gang device box might be referred to as a 3 × 2 × 3 inch box, a 3 inch × 2 inch × 3 inch deep box, or a trade size 3 × 2 × 3 box.

Trade sizes for construction material will not change. A 2 × 4 is really a name, not an actual dimension. A 2 × 4 will still be referred to as a 2 × 4. This is its trade size.

In this text, measurements directly related to the *NEC®* are given in both inch-pound and metric units. In some instances, only the inch-pound units are shown. This is particularly true for the examples of raceway and box fill calculations, load calculations for square foot areas, and on the plans (drawings).

Because the *NEC®* rounded off most metric conversion values, a computation using metrics results in a different answer when compared to the same computation done using inch-pounds. For example, load calculations for a residence are based on 3 volt-amperes per square foot or 33 volt-amperes per square meter.

- For a 40 × 50 foot dwelling: 3 VA × 40 ft × 50 ft = 6000 volt-amperes.

- In metrics, using the rounded off values in the *NEC®*: 33 VA × 12 m × 15 m = 5940 volt-amperes.

The difference is small, but nevertheless it is different.

To show calculations in both units throughout this text would be very difficult to understand, and would take up too much space. Calculations in either metrics or inch-pounds are in compliance with the *NEC®* 90.9(D). In 90.9(C)(3) we find that metric units are not required if the industry practice is to use inch-pound units.

It is interesting to note that the examples in Chapter 9 of the *NEC®* use inch-pound units, not metrics.

ACKNOWLEDGMENTS

The author would like to give special thanks to:

Allied Tube and Conduit
American Iron and Steel Institute (AISI)
Calculated Industries, Inc.
Carlon/Lamson and Sessions Company
Cooper B-Line Inc.
Cooper Bussman
Cooper Industries
Crouse-Hinds EMC
Pyrotenax USA Inc.
Square D Company
Taymac Corporation
Underwriters Laboratories, Inc.
Unity Manufacturing

The author and Thomson Delmar Learning would like to acknowledge the time and effort of the reviewers. Our thanks go to:

Michael Forester
Charles Trout

Applicable tables and section references are reprinted with permission from NFPA 70-2005, *National Electrical Code®* (*NEC®*), copyright © 2004–, National Fire Protection Association, Quincy, MA 02269. This reprinted material is not the complete and official position of the NFPA on the referenced subject, which is represented only by the standard in its entirety.

About the Author

Richard E. Loyd is a nationally known author and consultant specializing in the *National Electrical Code*® and the model building codes. He is president of his own firm, R & N Associates, located in Sun Lakes, Arizona. He and his wife, Nancy, travel throughout the country presenting seminars and speaking at industry-related conventions. He also serves as a code expert at 35 to 40 meetings per year at the International Association of Electrical Inspectors (IAEI) meetings throughout the United States. Mr. Loyd represented the National Electrical Manufacturers Association (NEMA) Section 5RN Steel Rigid Conduit and Tubing as an *NEC*® consultant from 1986 through April 1997. Mr. Loyd presently represents the Steel Tube Institute (STI) of North America as an *NEC*® and Model Codes consultant. He represents the American Iron and Steel Institute (AISI) on ANSI/NFPA 70 in the *National Electrical Code*® as a member of Code Committee Eight (8), which is the panel responsible for raceways, and Code Committee Five (5), which is the panel responsible for grounding. Mr. Loyd is actively involved doing forensic inspections and investigations on a consulting basis and serves as a special expert in matters related to the application of codes as they relate to installations and safety. He is a member of the Arizona Chapter of the Electrical Inspectors Association. He is chairman of the Electrical Section of the National Fire Protection Association (NFPA).

He is an active member in the International Association of Electrical Inspectors (IAEI); NFPA; the Institute of Electrical and Electronics Engineers (IEEE), where he is presently a member of the Power Systems Grounding Committee (Green Book); the International Conference of Building Officials (ICBO); the Southern Building Code Congress International (SBCCI); and the Building Officials and Code Administrators International (BOCA). Mr. Loyd is currently licensed as a master contractor/electrician in Arkansas

(license #1725) and Idaho (license #2077) and is an NBEE certified master electrician.

Mr. Loyd served as the chief electrical inspector and administrator for the State of Idaho and for the State of Arkansas. He has served as chairman of NFPA 79 "Electrical Standard for Industrial Machinery"; as a member of Underwriters Laboratories Advisory Electrical Council; as chairman for Educational Testing Service (ETS), a multistate electrical licensing advisory board; and he served as associate editor for Intertec Publications (*EC&M* Magazine) for 10 years. He is a master electrician and former electrical contractor. He has served as chairman of the National Board of Electrical Examiners (NBEE). He is an accredited instructor for licensing certification courses in Idaho, Oregon, and Wyoming and has taught basic electricity and *National Electrical Code*® classes for Boise State University (BSAU) in Boise, Idaho.

Chapter One

EXAMINATIONS AND NATIONAL TESTING ORGANIZATIONS

Examinations are an old phenomenon. They have been administered in many disciplines to evaluate competency in the field of specialty. Until recently they were prepared by others within the chosen field, based on the knowledge of each person preparing the examination. Today many examinations are professionally developed by state agencies specializing in examination preparation and by national testing organizations.

In recent years many local jurisdictions, such as city, county, and state, have dropped their electrical examinations in favor of using the national testing agencies that prepare and administer electrical examinations for those local entities. However, there are still many local jurisdictions developing and administering electrical examinations much in the same manner as has been done since the beginning of electrical examinations. These examinations are often prepared and administered by local electrical inspectors and are often developed with question material related to areas in which the inspector often finds violations or to areas in the *National Electrical Code®* (*NEC®*) that the inspector feels are most difficult to understand. Rarely do local examinations fully evaluate an entry-level craftsman regarding his or her competency in this new career field. National testing organizations, such as

Experior or the Southern Building Code Congress International (SBCCI), have carefully taken the advantages of localized testing methodology and incorporated the latest state-of-the-art technology to develop a test that is both fair to the taker and a good evaluation of the taker's knowledge in the specific career field. These testing organizations use those experienced in the trades in pools and in sit-down committees to develop a task analysis to determine what the examination will evaluate and then develop questions based on those tasks. These organizations strive to develop question criteria that are clear and not confusing, questions that have only one answer, and questions that an entry-level candidate entering the journeyman electrical field or the master contractor field should know as minimum standards.

These examinations are generally multiple-choice written examinations. The study arrangement in this book is designed to prepare the user to successfully pass these national examinations. However, it also should provide those preparing for local tests on these same subjects.

DEVELOPING AN ITEM (QUESTION) BANK

The development of examinations that effectively measure the minimum competency of an entry-level candidate is a process that uses many concepts. National testing firms—such as Experior (National Headquarters, 1260 Energy Lane, St. Paul, MN 55108), for construction trades multistate licensing, and Southern Building Code Congress International (SBCCI) (900 Montclair Road, Birmingham, AL 35213-1206)—have developed examinations that have become standards in the industry today. The first step in developing an examination is a task or job analysis. The task analysis must evaluate the ability, knowledge, and skill needed to perform the tasks related to the job in a way that will not endanger public health, safety, and welfare. The analysis must be relevant to the actual electrical practice in the field. The first step is then to establish the criteria, the subject areas of the trade that are most important and need to be tested. Two acceptable methods for determining these various job-related tasks would be (1) to survey licensed persons from the construction trades who perform these tasks in the field on a day-to-day basis, and (2) to assemble a group of competent experts in these fields to list the content outline and determine the degree of importance of each task to be evaluated by the examination and create a blueprint that shows which percent of the exam is designated for each subject task. This blueprint ties the test to the job performance, and the test outline shows the importance of each subject. When done properly, the task analysis will substantiate the validity of the exam.

The next task is to hold workshops or group meetings to develop an item (question) bank for a group of questions. This task of developing the item bank is normally performed by a group of interested experts working in workshops to develop the item bank based on the *NEC®* and other electrical related texts or reference books. As the item bank is developed, it is important that each item be clear, precise, and have only one correct answer. On completion of the item bank, there should be a pool of 500 to 800 questions. These questions are entered into the computer, each based on the task analysis for each job task and weighted according to the difficulty level of each question. The electrical board whose members come from contract states subscribing to these national tests then determines the passing scores, the item selection, and test form assembly. When this has been completed, the examination is ready for administration. Following the test administration to the candidates, a post-test examination analysis is then conducted in order to determine the effectiveness of the examination. In the post-test analysis, which is conducted following every exam, the effectiveness of each item is examined. Inadequate items are returned to the workshop to be rewritten, corrected, or eliminated from future exams. Each item is evaluated from developed criteria from previous exams, and the difficulty level is verified. Post-examination analysis provides critical and important information for future workshops and the validity of the exams. This analysis provides information that cannot be gained in any other manner. Information continues to be developed throughout the life of the item bank. This information is fed into the computer and continually proves and improves the validity of future examinations. The candidates themselves provide the most important information in the continuing development and improvement of the examination.

PREPARING FOR AN EXAMINATION

The first step in preparing for an examination is to obtain the examination information from the authority having jurisdiction (AHJ) to which you are about to make application for an examination. National testing firms furnish each AHJ a bulletin containing generic material that applies nationwide, with specific unique material related only to the jurisdiction in which you are about to take an examination. This bulletin contains information that must be read carefully about the eligibility and procedure for registering for the test. This includes your verification of experience and education requirements and usually requires an application fee. It also contains the deadlines that are important for submitting the application and fees and the examination dates. For instance, you may be required to have your application and fees submitted up to 60 days before the examination date. The fees vary from state to state. They generally require the application fee to be made to the jurisdiction and the exam administration fee to be made directly to the national testing firm. However, this procedure varies, and the instructions must be read carefully. Once your application has been approved, your name will be placed on the roster for the administration of the exam at the testing center that you have selected or has been assigned to you. This will generally be the testing center closest to your home address unless you have indicated that you would like to be tested in another location. A schedule of examination dates and locations for the year is included in the candidate bulletin or is available from the local jurisdiction. You will be mailed an admission letter. Do not lose the letter. It will be required when you arrive at the test center or test location to take the exam. If you fail to bring this letter, you will not be permitted to take the exam. Also, it is very important that you read this bulletin to determine what materials are required and what materials are allowed, usually the *NEC®* or handbook in the edition in which the test is developed.

> **Warning:** Verify which *NEC®* edition is being used for the examination (2002 or 2005). Check allowable reference books before studying for the examination. Some agencies permit only the softcover *NEC®*, others permit the *NFPA® Handbook* and other reference books. Verify the pencil type and the type of calculator allowed. Do not wait until test day. Be prepared.

Other electrical reference books may be permitted. You are usually required to bring your own pencils and a silent, handheld, nonprogrammable, battery-powered calculator. Some jurisdictions also require a hands-on practical test. If it is required in the area in which you are making application, the outline of that work should also appear in the bulletin. Read the outline carefully. The practical test can mean the difference between passing and failing the examination. One such jurisdiction required that the candidate bring conduit, a conduit bender, hacksaw, reaming tool, tape measure, and continuity or ohmmeter. Also contained in the candidate bulletin will be the requirements for candidates with physical disabilities. Documentation of the disability must be submitted at the time the application is made. Special considerations are generally permitted to accommodate religious or disability needs as permitted by the Americans with Disabilities Act (ADA).

Finally, the candidate bulletin contains a content outline; the number of questions in each content area; the study reference material that the test has been developed from, such as the *National Electrical Code*®, *Alternating Current Fundamentals* by Duff and Herman (Thomson Delmar Learning), *Direct Current Fundamentals* by Loper and Tedsen (Thomson Delmar Learning), *The American Electrician's Handbook*; and sample questions. It is suggested that you cover the answers and take this sample test in much the same manner as you would take the real examination. This will give you an idea of what the test will be like and will familiarize you with the format of the test if you are not already familiar with multiple-choice examinations. This is an important part of preparing for the electrical examination. Take the candidate bulletin seriously.

The next step in preparing for the examination is to study this book. The review questions at the end of each chapter are taken from the *NEC*® and not necessarily only from the content of each chapter. The *NEC*® is a reference document. As a reference document, it is necessary that you develop a strong proficiency in using the Table of Contents and the Index to be able to quickly locate the articles and sections of the *Code* you need. For each review question that you answer, you should not only research and find the correct answer, but you should also reference the *NEC*® section in which the answer was found. As you go through all the review material, take the sample tests throughout the document. This will enable you to pass the test the first time you take it. Anything less may prove disappointing.

> Each lesson is designed purposely to require the student to apply the entire *NEC*® text, and not just specific chapters or articles. It has been found that when studying for a timed, open-book examination, the student must gain proficiency in the Table of Contents, the Index, and the ability to move quickly from cover to cover in order to find the correct answer to each question in a timely fashion.

You should be aware that if you fail the examination, some jurisdictions require a waiting period of six months to one year before you can retake it.

The next section of this book contains a sample examination similar to the examination that you will take in the jurisdiction that you are studying for and similar to the examination given by many national testing organizations. This test should be taken in a quiet room. You should time yourself for three hours. The examination should be taken all at one time. To do otherwise will not give you a true evaluation of your present knowledge and of the areas in which you are weakest. Learning which areas are your weakest will enable you to concentrate your efforts and study time to gain the competency needed to pass your examination.

You may have purchased this book for various reasons. Some individuals study merely to improve their understanding in their chosen field, but most do so because they are required to do so to pass their performance evaluation on their job or to pass an evaluation in a state or local examination. The specific reason may be unimportant. The important thing is that you are about to undertake the study of electricity that, if done properly, will improve your competency in this field. The better you prepare, the better you will do on any evaluation or examination. Preparedness has more than one effect; for example, your IQ can fluctuate 30 to 40 points between any given dates.

The best way to improve individual performance is to reduce the anxiety level. To improve your study habits, consistent study time is important. You should allow yourself the same amount of time each day or each week to study this course. Consistent study time improves your retention.

This book and other Thomson Delmar Learning books related to this subject are designed to improve an individual's understanding of how to use the *NEC*® as a reference document and how to use the basic electrical formulas as applicable. They duplicate as nearly as possible the test format used by many national testing organizations and local jurisdictions. Whether you are preparing for an examination or just improving your skills, good study habits will give you the maximum efficient use of your time.

Basic Minimum Preparedness Tools

By the time test day arrives, you should have the achieved the following:

1. The *NEC®* is a reference book. Most examinations recognize this and are "open-book exams." However, it is still necessary that you are comfortable using it as a tool.

   • Be comfortable using the *NEC®* Table of Contents. Understand each chapter heading and where and how to find things quickly.

   • Be comfortable using the *NEC®* Index using key words.

   • Have a thorough understanding of *NEC®* Article 90—Introduction. This article gives you the *Code* purpose, what is covered and not covered, *Code* arrangement, enforcement, rules, and planning.

   • Have a thorough understanding of *NEC®* Article 100—Definitions. This article contains special definitions that are specific to this document, different from the definition you find in a standard dictionary, such as "bathroom," "accessible," and "location."

   • Have a thorough understanding of *NEC®* Article 110—Requirements for Electrical Installations. This article covers general requirements applicable to all installations.

   • Have a thorough understanding of *NEC®* Article 300—Wiring Methods. This article covers general wiring method requirements applicable to all types of wiring methods.

2. Basic electrical mathematical formulae are a must if you are to be successful.

   • voltage drop

   • Ohm's law

   • common alternating current (ac) formulae

If you have achieved this basic necessary knowledge and you have completed this book, you should have no problem passing any standard electrical exam.

Get a good night's sleep the day before the test.

ALTERNATING CURRENT		
To Find	Single Phase	Three Phase
Amperage when "HP" is known	$\dfrac{HP \times 746}{E \times \%EFF \times PF}$	$\dfrac{HP \times 746}{E \times \%EFF \times PF \times 1.73}$
Amperes when "KW" is known	$\dfrac{KW \times 1000}{E \times PF}$	$\dfrac{KW \times 1000}{E \times PF \times 1.73}$
Amperes when "KVA" is known	$\dfrac{KVA \times 1000}{E}$	$\dfrac{KVA \times 1000}{E \times 1.73}$
Kilowatts	$\dfrac{E \times I \times PF}{1000}$	$\dfrac{E \times I \times PF \times 1.73}{1000}$
Kilovolt-Amperes "KVA"	$\dfrac{E \times I}{1000}$	$\dfrac{E \times I \times 1.73}{1000}$
Horsepower	$\dfrac{E \times \%EFF \times PF}{746}$	$\dfrac{E \times I \times \%EFF \times PF \times 1.73}{746}$

Percent Efficiency = $\dfrac{\text{Output}}{\text{Input}}$

PF = $\dfrac{\text{Power (Watts)}}{\text{Apparent Power}}$ = $\dfrac{KW}{KVA}$

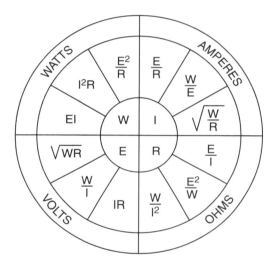

Test Day

You should have the following to take with you to the examination. Check your Candidate Information Bulletin to confirm allowable materials.

   • Admission mailer (letter).

   • Photo I.D., such as a driver's license.

   • *NEC®* book (current adopted edition) and any other approved reference material that you feel will benefit you in taking the test. ***Note:*** Your reference material will be checked for handwritten notes and other markings not acceptable to the jurisdiction in which you are testing.

   • A silent, nonprinting, nonprogrammable calculator that is not capable of alphabetic entry.

   • Two No. 2 pencils.

The time has finally arrived for you to take your examination. Be sure you allow plenty of time to travel to the testing location; it is important to arrive on time. Some testing agencies will not permit admission if you arrive late. On the other hand, do not arrive too early. Discussing the test with other examinees may raise your anxiety level. Nervousness is contagious! Be sure you have your receipt for your application and your entrance ticket. Be sure you have all materials required by the jurisdiction or testing agency, including study books and a handheld calculator.

When a test is multiple choice, you generally have four alternative answers; therefore, the odds are four to one that you can get the right answer. If you can eliminate two of the alternatives, your chances are now increased to 50/50. This is extremely beneficial when referencing the *Code*, as you can quickly eliminate half of the possibilities leaving you only two to research.

A lighting and branch-circuit panelboard is installed in a dwelling unit bathroom. Which of the following answers is correct?
1. It is not a *NEC®* violation.
2. It is permitted, but it must be suitable for wet locations.
3. The location of panelboards is not covered by the *NEC®*.
4. It is a *NEC®* violation.

We can quickly eliminate 2 and 3. The answer is 4. The *NEC®* 240.24(E) prohibits installing overcurrent devices in dwelling unit bathrooms.

Read the directions carefully. Many mistakes are made merely because of misunderstanding. If you have any questions, discuss them with the proctor before the test time starts. Any discussions after the test starts will take away from the time you have to answer the questions. Allow yourself ample time for each question. Answer the question first before you check any alternatives. This way you can evaluate your answer against any alternatives that you may find in the *NEC®*. Do not spend too much time on any one question. If you find the question extremely difficult and you are unsure of the answer or cannot find the answer in the *NEC®*, skip that question and go back to it only if you have time at the end of the test.

You should never spend more than 3½ minutes per question. Your first choice is usually best. Read the questions very carefully. Make sure you know key words that might change the meaning of the question. Note any negatives, such as "Which of the following

are not, or shall not be permitted?" It may be beneficial to underline key words as you read the question. This has a tendency to channel your thoughts in the right direction.

Sample Question

Tip: Break the question down and get to the root question.

The bathroom in a custom home includes a vanity with double sinks (basins) located on one side of the bathroom. There is an outlet located adjacent to each sink (basin) with ground-fault circuit-interrupter protection (GFCI). On the other side of the bathroom, just opposite to and about 8 feet away, is a makeup vanity (without a sink [basin]) with a receptacle outlet located at the same height as the GFCI outlets across the room.

Would this outlet above this vanity be required to also have GFCI protection?

Answer:

The key words are: bathroom
home (dwelling)
GFCI

The question is, Would an outlet not adjacent to a sink in a bathroom need GFCI protection? The answer is Yes, all receptacles located in a residential bathroom are required to be protected with ground-fault circuit-interrupter (GFCI) protection. The reference is *NEC®* 210.8(A)(1).

This question could be time-consuming because it causes you to look at the location in relation to the sink. *NEC®* 210.52(D) requires a receptacle to be located within 3 feet (900 mm) of the outside edge of each basin (sink). However, the question does not ask that! The question asks whether another receptacle located outside this dimension needs GFCI protection.

Summary

- Dress properly; wear comfortable shoes and loose-fitting clothing. The test facility should be approximately 70°F. Too many clothes or uncomfortable clothes have a tendency to make you drowsy and concentration difficult.

- Plan to arrive at the test site 20 to 30 minutes early. Anxiety will reduce your effectiveness. Do not be late or too early. Make sure ahead of time that you know where the test site is located and whether parking is available.

- Be sure you have your acceptance letter and proper I.D. Verify ahead of time so that you have the identification materials required by the testing agency.

- Do not forget your *NEC*® book, reference books, calculator, and other acceptable/required materials. Pack all these materials the night before in a backpack or briefcase.

- After you have answered all the questions, recheck your work.

1. Make sure that you have not made clerical mistakes.
2. Make sure that your answers are recorded on the correct line on the answer sheet.
3. Go over the extremely difficult questions.
4. Regarding questions on which you know you have guessed—you may want to mark them in some way so that if there is time at the end of the test, you can do additional research.

Good luck!

SAMPLE EXAMINATION
Examination Instructions:

For a positive evaluation of your knowledge and preparation awareness, you must:

1. Locate yourself in a quiet atmosphere (room by yourself).
2. Have with you at least two sharp No. 2 pencils, the 2005 *NEC*®, and a handheld calculator.
3. Time yourself (three hours) with no interruptions.
4. After the test is complete, grade yourself honestly and concentrate your studies on those sections of the *NEC*® in which you missed the questions.

   Caution: Do not just look up the correct answers because the questions in this examination are only an exercise and not actual test questions. Therefore, it is important that you be able to quickly find answers from throughout the *NEC*®.

Directions: Each question is followed by four suggested answers: A, B, C, and D. In each case, select the **one** that best answers the question.

1. Liquidtight flexible metal conduit can be used in which of the following locations?
   A. in areas that are both exposed and concealed
   B. in areas that are subject to physical damage
   C. in connection areas for gasoline dispensing pumps
   D. in areas where ambient temperature is to be greater than 200°C

   Answer: _____   Reference: _____

2. The maximum number of quarter bends allowed in one run of nonmetallic rigid conduit is
   A. 2.
   B. 4.
   C. 6.
   D. 8.

   Answer: _____   Reference: _____

3. The maximum size of flexible metallic tubing that can be used in any construction or installation is
   A. trade size ⅜ (12).
   B. trade size ½ (16).
   C. trade size ¾ (21).
   D. trade size 1 (27).

   Answer: _____   Reference: _____

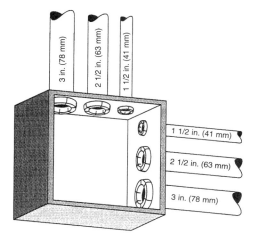

4. The size of the pull box in the diagram above should not be less than
    A. 12 inches × 14 inches (300 mm × 350 mm).
    B. 16 inches × 18 inches (400 mm × 450 mm).
    C. 22 inches × 22 inches (560 mm × 560 mm).
    D. 24 inches × 24 inches (600 mm × 600 mm).

    Answer: _____   Reference: _____

5. A metal raceway system installed in the washing area of a car wash must be spaced at least how far from the walls?
    A. ⅛ inch (3 mm)
    B. ¼ inch (6 mm)
    C. ¾ inch (19 mm)
    D. 1 inch (25 mm)

    Answer: _____   Reference: _____

6. Unless otherwise indicated, busways should be supported at intervals not to exceed how many feet?
    A. 3 (900 mm)
    B. 5 (1.5 m)
    C. 7 (2.1 m)
    D. 10 (3.0 m)

    Answer: _____   Reference: _____

7. Which of the following percentages of conduit fill should be used when four Type THW conductors are being installed?
    A. 55 percent
    B. 53 percent
    C. 40 percent
    D. 31 percent

    Answer: _____   Reference: _____

8. Flexible 18 AWG supply cords of listed appliances are considered adequately protected if the circuit overcurrent device is set at a maximum value of how many amperes?
A. 15
B. 20
C. 30
D. 40

Answer: _____   Reference: _____

9. All of the following sizes of solid aluminum conductors shall be made of an aluminum alloy except
A. 8 AWG.
B. 10 AWG.
C. 12 AWG.
D. 14 AWG.

Answer: _____   Reference: _____

10. An insulated bushing is required to be used on a raceway entering a cabinet if the ungrounded conductors entering the raceway are at least
A. 10 AWG.
B. 8 AWG.
C. 6 AWG.
D. 4 AWG.

Answer: _____   Reference: _____

11. The allowable distance between supports of nonmetallic-sheathed cable installed in an on-site, constructed, one-family dwelling is a maximum of how many feet?
A. 2½ (750 mm)
B. 3 (900 mm)
C. 4 (1.2 m)
D. 4½ (1.4 m)

Answer: _____   Reference: _____

12. It is permissible to use 4 THW CU AWG for service in dwelling units to a maximum load of
A. 100 amperes.
B. 95 amperes.
C. 70 amperes.
D. 65 amperes.

Answer: _____   Reference: _____

13. Where conductors of less than 600 volts emerge from the ground, they must be protected by enclosures or raceways that extend from below grade to a point how many feet above the finish grade?
A. 5 (1.5 m)
B. 6 (1.8 m)
C. 7 (2.1 m)
D. 8 (2.4 m)

Answer: _____   Reference: _____

14. The service rating for a residence that uses 2 AWG THW copper service-entrance conductors is
    A. 95 amperes.
    B. 110 amperes.
    C. 115 amperes.
    D. 125 amperes.

    Answer: _____   Reference: _____

15. Surge arrester grounding conductors that run in metal enclosures should be
    A. bare.
    B. insulated.
    C. bonded on one end of the enclosure only.
    D. bonded at both ends of the enclosure.

    Answer: _____   Reference: _____

16. The number of overcurrent devices in a single cabinet of a lighting and appliance panelboard should not exceed
    A. 30.
    B. 36.
    C. 42.
    D. 48.

    Answer: _____   Reference: _____

17. When it is impractical to locate a service head directly above the point of attachment, the maximum allowable placement distance from the point of attachment is how many feet?
    A. 1 (300 mm)
    B. 2 (600 mm)
    C. 3 (900 mm)
    D. 4 (1.2 m)

    Answer: _____   Reference: _____

18. An underground service to a small, controlled water heater is to be installed. The single branch-circuit used for this service requires that copper conductors be at least
    A. 12 AWG.
    B. 10 AWG.
    C. 8 AWG.
    D. 6 AWG.

    Answer: _____   Reference: _____

19. Service conductors can have less than 3 feet (900 mm) of clearance from which of the following?
    A. doors
    B. tops of windows
    C. fire escapes
    D. porches

    Answer: _____   Reference: _____

20. Service conductors that pass over rooftops should have a vertical clearance of not less than how many feet?
    A. 3 (900 mm)
    B. 7 (2.1 m)
    C. 8 (2.4 m)
    D. 10 (3.0 m)

    Answer: _____   Reference: _____

21. The interrupting rating shall be marked on all circuit breakers rated more than
    A. 1000 amperes.
    B. 2000 amperes.
    C. 2500 amperes.
    D. 5000 amperes.

    Answer: _____   Reference: _____

22. Illumination is mandatory for service equipment or panelboards in a dwelling unit if the service to the unit exceeds how many amperes?
    A. 100
    B. 200
    C. 400
    D. required for all services regardless of amperage

    Answer: _____   Reference: _____

23. Ground-fault protection is required for _____-ampere, 480-volt, 4-wire, 3-phase solidly grounded wye services.
    A. 800
    B. 1000
    C. 1200
    D. above 2000

    Answer: _____   Reference: _____

24. Impedance is present in which of the following circuits?
    A. ac only
    B. dc only
    C. resistance only
    D. both ac and dc

    Answer: _____   Reference: _____

25. Excluding fuses and exceptions, how many overcurrent protection devices, such as trip coils, relays, or thermal cutouts, are required on a 3-phase motor?
    A. 5
    B. 3
    C. 2
    D. 0

    Answer: _____   Reference: _____

26. The attachment plug and receptacle can be used as the controller for a portable motor that has a maximum horsepower rating of _____ horsepower.
    A. ⅓
    B. ½
    C. 1
    D. 2

27. Snap switches can be grouped or ganged in outlet boxes if voltages between adjacent switches do not exceed how many volts?
    A. 200
    B. 300
    C. 400
    D. 500

    Answer: _____   Reference: _____

28. Each doorway that leads into a transformer vault from the building interior should have a tight-fitting door with a minimum fire rating of how many hours?
    A. 1
    B. 2
    C. 3
    D. 5

    Answer: _____   Reference: _____

29. The conduit run between warm and cold locations should be
    A. sealed.
    B. sleeved.
    C. insulated.
    D. provided with drains.

    Answer: _____   Reference: _____

30. Except for points of support, recessed portions of luminaire (lighting fixture) enclosures should be spaced at least how many inches from combustible material?
    A. ¼ (6 mm)
    B. ½ (13 mm)
    C. 1 (25 mm)
    D. 3 (75 mm)

    Answer: _____   Reference: _____

31. Surface-mounted fluorescent luminaires (lighting fixtures) that contain a ballast and are to be installed on combustible, low-density cellulose fiberboard should be spaced not less than how many inches from the surface of the fiberboard?
    A. ½ (13 mm)
    B. 1 (25 mm)
    C. 1½ (38 mm)
    D. 2 (50 mm)

    Answer: _____   Reference: _____

32. The minimum amount of on-site fuel for an emergency generator should be sufficient for not less than
    A. 2 hours at full demand.
    B. 4 hours at 75 percent demand.
    C. 6 hours at 50 percent demand.
    D. 8 hours at 25 percent demand.

    Answer: _____   Reference: _____

33. Which of the following statements about dimmers installed in theaters is (are) true?
    I. Dimmers installed in ungrounded conductors must have overcurrent protection not greater than 125 percent of the dimmer rating.
    II. The circuit-supplying autotransformer type dimmers must not exceed 150 volts between conductors.
    A. I only
    B. II only
    C. both I and II
    D. neither I nor II

    Answer: _____   Reference: _____

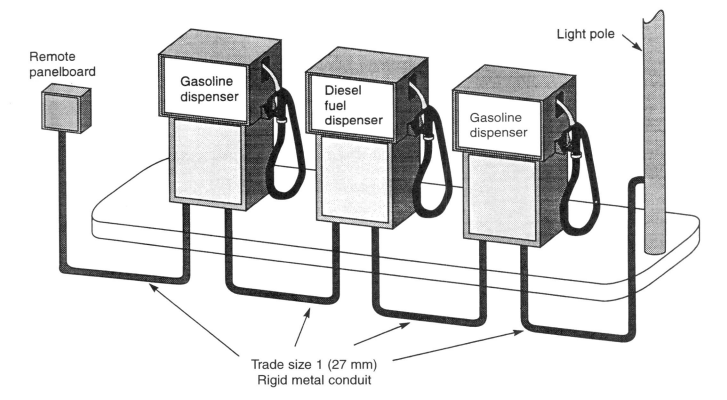

34. In the diagram above, how many conduit seals are required?
    A. 2
    B. 4
    C. 6
    D. 8

    Answer: _____   Reference: _____

35. An area must be classed as a Class II hazardous location if it contains which of the following?
    A. combustible dust
    B. flammable gases
    C. ignitable liquid
    D. ignitable vapors

    Answer: _____   Reference: _____

36. If the wiring is to be placed 18 inches (450 mm) above the floor, which of the following wiring methods is prohibited for a commercial automotive repair shop?
    A. metal-clad cable
    B. electrical metallic tubing
    C. rigid nonmetallic conduit
    D. nonmetallic-sheathed cable

    Answer: _____   Reference: _____

37. The minimum number of receptacles in a patient bed location of a hospital's general care area should be
    A. 1 duplex.
    B. 2 duplex.
    C. 3 duplex.
    D. 3 single.

    Answer: _____   Reference: _____

38. Which of the following statements about lightning protection is (are) true?
    I. If lightning protection is required for an irrigation machine, a driven ground rod should be connected to the machine at a stationary point.
    II. Lightning rods should be spaced at least 6 feet (1.8 m) away from noncurrent-carrying metal parts of electrical equipment.
    A. I only
    B. II only
    C. both I and II
    D. neither I nor II

    Answer: _____   Reference: _____

39. Which of the following will effectively ground a 230-volt, single-phase, residential air-conditioning unit that replaces an older unit supplied by a 2-wire, 230-volt circuit?
    I. a new circuit containing a grounding conductor
    II. an 8-foot (2.5 m) ground rod installed at the unit
    A. I only
    B. II only
    C. both I and II
    D. neither I nor II

    Answer: _____   Reference: _____

40. When connected to a rod, pipe, or plate grounding electrode, grounding electrode copper conductors need not be sized larger than
    A. 8 AWG.
    B. 6 AWG.
    C. 4 AWG.
    D. 2 AWG.

    Answer: _____   Reference: _____

41. Which of the following is the most acceptable grounding electrode?
    A. a driven steel approved ground rod
    B. well casing
    C. building foundation steel
    D. an underground cold water metal piping system

    Answer: _____   Reference: _____

42. Which of the following terms can be used in place of "resistance" in the phrase "resistance is to ohms"?
    A. reactance
    B. inductance
    C. impedance
    D. capacitance

    Answer: _____   Reference: _____

43. The *NEC®* requires a minimum of which of the following for the kitchen small-appliance load in dwelling units?
    A. two 20-ampere circuits
    B. two 15-ampere circuits
    C. one 20-ampere circuit
    D. one 15-ampere circuit

    Answer: _____   Reference: _____

44. Ground-fault protection of equipment is a requirement for solidly grounded wye services that are rated at more than 150 volts-to-ground but do not exceed 600 volts, phase-to-phase, for each service disconnecting means rated at 1000 amperes or more. The maximum amperage for this ground-fault protection relay should be set at
    A. 1000 amperes.
    B. 1200 amperes.
    C. 1500 amperes.
    D. 2000 amperes.

    Answer: _____   Reference: _____

Number	Size (kcmil)	Type	Function
2	500	THW	Phases A and B
1	400	THW	Neutral
1	300	THW	Phase C
1	3 AWG	THW	Ground Conductor

45. The minimum size intermediate metal conduit required for installation of the conductors above should not be less than
    A. trade size 2 (53).
    B. trade size 2½ (63).
    C. trade size 3 (78).
    D. trade size 3½ (91).

    Answer: _____   Reference: _____

46. The combined load of several 240-volt, fixed space heaters on a 20-ampere circuit should not exceed how many kilowatts?
    A. 2.4
    B. 2.6
    C. 3.8
    D. 4.8

    Answer: _____ Reference: _____

47. If three resistors with values of 5, 10, and 15 ohms, respectively, are connected in parallel, the combined resistance of the units will be
    A. 1.63 ohms.
    B. 2.73 ohms.
    C. 20.0 ohms.
    D. 30.0 ohms.

    Answer: _____ Reference: _____

48. The *NEC*® covers the installation of electrical conductors and equipment in all of the following locations except
    A. public and private buildings.
    B. floating dwelling units.
    C. buildings used by a utility for warehousing.
    D. centers of transmission and distribution of electrical energy.

    Answer: _____ Reference: _____

49. The most recent editions of the *NEC*® are revised every
    A. year.
    B. 2 years.
    C. 3 years.
    D. 4 years.

    Answer: _____ Reference: _____

50. A capacitor is located indoors. It must be enclosed in a vault if it contains more than a minimum of _____ gallon(s) of flammable liquid.
    A. 1
    B. 2
    C. 3
    D. 5

    Answer: _____ Reference: _____

Chapter Two

OBJECTIVES

After completing this chapter, you should be able to:

- Identify the largest organizations that impact the electrical industry and give a brief outline of what each does and how they each impact the electrical industry (e.g., NFPA, NRTL)
- Identify agencies responsible for enforcement and regulation of the *NEC®* and its amendments, or other electrical codes
- Explain the basic authority of the *NEC®*
- Explain how the *NEC®* is organized and developed and identify segments of the industry covered by the *NEC®* and those that are not
- Explain how testing laboratories impact code enforcement

NATIONAL FIRE PROTECTION ASSOCIATION STANDARDS

The National Fire Protection Association (NFPA) has acted as the sponsor of the *National Electrical Code®* (*NEC®*) as well as many other safety standards. The most widely used electrical code in the world is the *NEC®*. The official designation for the *NEC®* is ANSI/NFPA 70.

NATIONAL ELECTRICAL CODE® (*NEC®*)

ANSI/NFPA 70, the *National Electrical Code®* (*NEC®*), was first developed about 100 years ago by interested industry and governmental authorities to provide a standard for the safe use of electricity. It is now revised and updated every three years. It was first developed to provide the regulations necessary for safe installations and to provide the practical safeguards of persons and property from the hazards arising from the use of electricity. It still provides these regulations and safeguards to the industry today. The *NEC®* is developed by 20 different Code-Making Panels composed entirely of volunteers. These volunteers come from industry,

testing laboratories, inspection agencies, engineering, users, government, and the electrical utilities. Suggestions for the content come from various sources, including individuals like you. Anyone can submit a proposal or make a comment on a proposal submitted by another individual.

For more specifics on the *NEC®* process, you can contact the National Fire Protection Association, One Batterymarch Park, Quincy, MA, 02269-9101. Request their free booklet "The NFPA Standards Making System."

The *NEC®* is advisory as far as NFPA and ANSI are concerned but is offered for use in law and regulations in the interest of life and property protection. The name *National Electrical Code®* might lead one to believe that this document is developed by the federal government. This is not so; the *NEC®* only recognizes uses of products. It approves nothing. The *NEC®* has no legal standing until it has been adopted by the *authority having jurisdiction* (AHJ) (see *NEC®* definition Article 100), usually a governmental entity. Therefore, we must first check with the local electrical inspector to see which edition of the *NEC®* has been adopted. Compliance within and proper maintenance will result in an installation essentially free from hazard but not necessarily efficient, convenient, or adequate for good service or future expansion of electrical use. The *NEC®* is not an instruction manual for untrained persons, nor is it a design specification. However, it does offer design guidelines.

NEC® 90.2 defines the scope of this document. It is intended to cover all electrical conductors and equipment within or on public and private buildings or other structures, including mobile homes, recreational vehicles, and floating buildings, and other premises such as yards, carnival, parking, and other lots, and industrial substations. Installations of conductors and equipment that connect to the supply of electricity, other installations of outside conductors on the premises, and installations of optical fiber cables are clearly covered by the *NEC®*. The *NEC®* is not intended to cover installations on ships, watercraft, railway rolling stock, automotive vehicles, underground mines, and surface mobile mining equipment. The *NEC®* is not intended to cover installations governed by the utilities, such as communication equipment, transmission, generation, and distribution installations on right-of-way.

Note: For the complete list of the exemptions and coverage, see 2005 *NEC®* 90.2. Mandatory rules are characterized by the word "shall." Explanatory rules are in the form of fine print notes (FPN). All tables and footnotes are a part of the mandatory language. Material text extracted from other NFPA standards and documents shall not be compromised or violated. Editing is permitted for adapting the extracted text into *NEC®* style. A complete list of all NFPA documents referenced can be found in Annex A. New revisions are identified by a vertical line on the margin where inserted.

To use the *NEC®*, one must first have a thorough understanding of Article 90—The Introduction; Article 100—Definitions; Article 110—Requirements for Electrical Installations; and Article 300—Wiring Methods. The rest of the *Code* can be referred to on an as-needed basis.

New for the 2005 *National Electrical Code®*

The 2005 *NEC®* has some notable changes but not as many or as significant as have been adopted in past editions. Some of the major changes that appear in the 2005 *NEC®* are noted here.

Article 80 has been relocated as a new addendum G. This was a logical move because Article 80 was never intended to be enforceable unless adopted separately.

Handhole enclosures have been added to Article 314. These are not new. They have been used by utilities for many years but are now being recognized in the *NEC®*. They are generally access enclosures without a bottom for access to splices for direct buried conductors.

A new Article 353 for high-density polyethylene (HDPE) conduit. This is a limited-use product that is only permitted outside underground.

Article 366—Auxiliary Gutters and Article 368—Busways have been rewritten and renumbered to conform to the numbering style in Chapter 3.

A new Article 409—Industrial Control Panels, has been added. This covers many types of contactors and control panels commonly fabricated to UL 508.

Adjustable speed drives (variable speed ac) have been added to Article 430. These have become very popular in all industries and needed to be added to the *NEC®*.

A new Article 506 is added to cover the Zone alternative system for hazardous dusts. This article will look similar to Article 505, which was added two editions earlier for hazardous gases and vapors.

Temporary Installations have been relocated from Article 305 in the 1999 *NEC®* to Article 527 in the 2002 *NEC®*, and in the 2005 *NEC®* will be renumbered as Article 590. There are no significant changes to this article.

A new Article 682 has been added to cover bodies of water not covered by Article 680. This will include fish farms and similar bodies of water.

Table 344.24, covering the bending radius allowed for raceways, is relocated to Chapter 9 as a new Table 2.

These are the major differences, but remember that there are hundreds of editorial and substantive changes that are enforceable. As you prepare for your examination, check all applicable *Code* sections so that your answers are accurate.

Article 90—Introduction

Article 90 contains the scope of the *NEC®* and is enforceable. This article contains information that applies to every installation. Every electrician should read this information carefully and understand it.

Chapter 1—General

Chapter 1 includes Article 100—Definitions. These definitions are unique to this document and apply where the term is used in the *NEC®*. Article 110 covers the general requirements for all electrical installations and contains information that applies to every installation. Every electrician should read this information carefully and understand it completely. These requirements apply to every installation regardless of how large or small it is.

Chapter 2—Wiring and Protection

Chapter 2 covers Article 200 through Article 285. These are general requirements that apply to all installations from Article 200, the use and identification of the grounded or neutral conductor, to a new Article 285 covering transient voltage surge suppressors (TVSS). Although the requirements in this chapter are applicable to all installations, the electrician should

refer to the material because there are too many requirements to remember.

Chapter 3—Wiring Methods

Chapter 3 covers Article 300 through Article 398. This
chapter was completely reorganized in the 2002 cycle
and renumbered. Temporary Wiring has been moved
to 590 and renamed Temporary Installations.

Article 300 covers the general wiring methods that
apply to all of the wiring methods in Chapter 3. Every
electrician needs to read this article carefully and understand it because these requirements apply to every installation regardless of how large or small it is. The balance
of this chapter is used where applicable. The electrician should use it as reference material because there
are too many requirements for anyone to remember.

Chapter 4—Equipment for General Use

Chapter 4 covers Article 400 through Article 490. This
chapter covers the general requirements for equipment
for all installations. Article 404—Switches and Article
408—Switchboards and Panelboards have been moved
to Chapter 4. Article 410 now covers only luminaires
(lighting fixtures), lampholders, and lamps. The
requirements for receptacles, cord connectors, and attachment plugs (caps) are covered in Article 406. A
new Article 409 has been added to cover industrial
control panels. Although it covers general equipment
such as appliances and motors that are used on all
installations, the electrician should use this chapter as
reference material because there are too many requirements for anyone to remember.

Chapter 5–Special Occupancies

Chapter 5 covers Articles 500 through 555. These
chapters do not apply generally—they modify or amend
Chapters 1 through 4. They are only applicable in the
types of occupancies or portions of buildings or structures that contain areas with special occupancies, such
as an area within another building that contains hazardous materials. However, Article 305—Temporary
Wiring has been relocated to Chapter 5 as Article 590.
The electrician should use Chapter 5 as reference
material because there are too many requirements for
anyone to remember.

Chapter 6—Special Equipment

Chapter 6 covers Article 600 through Article 695.
These chapters do not apply generally—they modify
or amend Chapters 1 through 4. They are only appli-

2005 NEC®	General Title
300	General Wiring Methods
310	Conductors for General Wiring
312	Cabinets, Cutout Boxes and Meter Socket Enclosures
314	Outlet, Device, Pull and Junction Boxes, Conduit Bodies and Fittings
320	Armored Cable: Type AC
322	Flat Cable Assemblies: Type FC
324	Flat Conductor Cable: Type FCC
326	Integrated Gas Spacer Cable: Type IGS
328	Medium Voltage Cable: Type MV
330	Metal-Clad Cable: Type MC
332	Mineral-Insulated Metal-Sheathed Cable: Type MI
334	Nonmetallic-Sheathed Cable: Types NM, NMC, and NMS
336	Power and Control Tray Cable: Type TC
338	Service-Entrance Cable: Types SE and USE
340	Underground Feeder and Branch-Circuit Cable: Type UF
342	Intermediate Metal Conduit: Type IMC
344	Rigid Metal Conduit: Type RMC
348	Flexible Metal Conduit: Type FMC
350	Liquidtight Flexible Metal Conduit:Type LFMC
352	Rigid Nonmetallic Conduit: Type RNC
353	High Density Polyethylene Conduit: Type HDPE
354	Nonmetallic Underground Conduit with Conductors: Type NUCC
356	Liquidtight Flexible Nonmetallic Conduit: Type LFNC
358	Electrical Metallic Tubing: Type EMT
360	Flexible Metallic Tubing: Type FMT
362	Electrical Nonmetallic Tubing: Type ENT
366	Auxiliary Gutters
368	Busways
370	Cablebus
372	Cellular Concrete Floor Raceways
374	Cellular Metal Floor Raceways
376	Metal Wireways
378	Nonmetallic Wireways
380	Multioutlet Assembly
382	Nonmetallic Extensions
384	Strut-Type Channel Raceway
386	Surface Metal Raceways
388	Surface Nonmetallic Raceways
390	Underfloor Raceways
392	Cable Trays
394	Concealed Knob-and-Tube Wiring
396	Messenger Supported Wiring
398	Opening Wiring on Insulators
404	Switches
406	Receptacles, Cord Connectors, and Attachment Plugs (Caps)
408	Switchboards and Panelboards
409	Industrial Control Panels
590	Temporary Installations

cable where the special equipment covered, such as a sign on a building or a swimming pool, is installed. The electrician should use Chapter 6 as reference material because there are too many requirements for anyone to remember.

Chapter 7—Special Conditions

Chapter 7 covers Article 700 through Article 780. These chapters do not apply generally—they modify or amend Chapters 1 through 4. They are only applicable where the special systems covered, such as limited voltage or emergency systems, are installed. The electrician should use Chapter 7 as reference material when encountering a system covered in this chapter because there are too many requirements for anyone to remember.

Chapter 8—Communications Systems

This chapter covers Article 800, Article 810, Article 820, and Article 830. Chapter 8 stands alone. It is not subject to the requirements of Chapters 1 through 7 unless the requirements are specifically referenced in Chapter 8. These chapters do contain many references, such as requirements for grounding. The electrician should use Chapter 8 only as reference material when installing a communication system. There are too many requirements for anyone to remember.

Chapter 9–Tables

These tables are applicable where referenced elsewhere in the *Code*. Table 344.24 has been relocated as Chapter 9, Table 2.

Annex A–G

Annex A through G (formerly "Appendix") contains informational material that is not mandatory or enforceable. Article 80 in the 2002 *NEC*® has been relocated as a new Addendum G.

The *NEC*® has a new way of numbering the parts to each article and the sections and subsections. The parts, which formerly appeared as A, B, C, and so on, now appear as Roman numerals such as Part I, Part II, Part III, and so on. The sections and subsections have a new look; they are formatted differently. The hyphen between the article number and section designation has been changed to a dot (.). The articles in the *NEC*® subdivide the chapters by a specific subject such as services, branch circuits, grounding, transformers, signs, and the like. The parts are subdivided by sections, and they may be subdivided by subsections or lists.

Example	
Chapter:	**Chapter 2—Wiring and Protection**
Article:	**Article 250—Grounding**
Part:	**IV. Conductors**
Section:	**250.119** Identification of Equipment Grounding Conductors. Unless required.
Level 1:	**(A)** Conductors larger than 6 AWG.
List item:	(1) Stripping the insulation or covering from the entire exposed length . . .
List item:	(2) Coloring the exposed insulation or covering green . . .
List item:	(3) Marking the exposed insulation or covering . . .
Level 2:	**(B)** Multiconductor Cable.
List item:	(1) Stripping the insulation . . .
List item:	(2) Coloring the exposed insulation . . .
List item:	(3) Marking the exposed insulation . . .

LOCAL CODES AND REQUIREMENTS

Although most municipalities, countries, and states adopt the *NEC*®, they may not have adopted the latest edition, and in some cases the authority having jurisdiction (AHJ) may be using an older edition and not the 2005 edition. Most jurisdictions make amendments to the *NEC*® or add local requirements. These may be based on environmental conditions, fire safety concerns, or other local experience. An example of one common local amendment is that all commercial buildings be wired in metal raceways (Rigid, IMC, or EMT). The *NEC*® generally does not differentiate between wiring methods in residential, commercial, or industrial installations; however, many local jurisdictions do. Metal wiring methods, especially in fire zones, are another common amendment. Some major cities have developed their own electrical code—for example, Los Angeles, New York, Chicago, and several metropolitan areas. In addition to the *NEC*® and all local amendments, the designer and installer must also comply with the local electrical utility rules. Most utilities have specific requirements for installing the service to the structure. There have been many unhappy designers and installers who have learned about special jurisdictional requirements after making the installation, thus incurring costly corrections at their own expense.

If you are preparing for a state or city examination, you must check with the AHJ and see exactly what the examination content is based on. If it is a locally developed examination, the content may vary widely. If it is a nationally developed examination, it will generally be based on the latest edition of the *NEC®* (2005 *NEC®*). However, some jurisdictions have supplemental examinations that cover their unique amendments, and in some jurisdictions a practical hands-on examination is given in addition to the written portion. (See Chapter One of this book for information related to some of these unique requirements.)

APPROVED TESTING LABORATORIES

Underwriters Laboratories (UL) has long been the major product testing laboratory in the electrical industry. In addition to the testing, UL is a longtime developer of product standards. By producing these standards and contracting for follow-up service after undergoing a listing procedure, a manufacturer is authorized to apply the UL label or to mark their product. Underwriters Laboratories is not the only testing laboratory evaluating electrical products. In the last decade, numerous other electrical testing laboratories have arrived and are being officially recognized; for example, Electrical Testing Laboratories (ETL), Applied Research Laboratories (APL), and Canadian Standards (CSA). Some jurisdictions evaluate and approve laboratories; others accept them based on reputation. OSHA is now evaluating and approving testing laboratories. It is the responsibility of the entity responsible for specifying the materials to verify that the product has been evaluated by a testing laboratory acceptable by the AHJ where the installation is being made. It is the installer's responsibility that the product is installed in accordance with the products listing (*NEC®* 110.3(B)).

QUESTION REVIEW

> Each lesson is designed purposely to require the student to apply the entire *NEC®* text, and not specific chapters or articles. It has been found that when studying for a timed, open-book examination, the student must gain proficiency in the Table of Contents, the Index, and the ability to move quickly from cover to cover to find the correct answer to each question in a timely fashion.

1. Is liquidtight flexible nonmetallic conduit permitted to be used in circuits in excess of 600 volts?

   Answer: _____

   _____

   _____

   _____

   Reference: _____

2. Who is the authority having jurisdiction (AHJ)?

   Answer: _____

   _____

   _____

   _____

   Reference: _____

3. Name four types of installations not covered by the *NEC®*.

   Answer: _____

   _____

   _____

   _____

   Reference: _____

4. How is explanatory information characterized in the *NEC®*?

   Answer: _____

   _____

   _____

   _____

   Reference: _____

5. What section of the *NEC®* requires equipment to be installed in accordance with manufacturer's instructions?

Answer: _____

_____

_____

_____

Reference: _____

6. When does the *NEC®* become the legal document?

Answer: _____

_____

_____

_____

Reference: _____

7. Are all installations by utilities exempt from the *NEC®*? Explain.

Answer: _____

_____

_____

_____

Reference: _____

8. What does the vertical line in the margin of the *NEC®* indicate?

Answer: _____

_____

_____

_____

Reference: _____

9. What is the latest amended edition of the *NEC®* used in any jurisdiction? When was the latest edition of the *NEC®* published?

Answer: _____

_____

_____

_____

Reference: _____

10. Who set up the first meeting that resulted in the *NEC®*? When and where was it held?

Answer: _____

_____

_____

_____

Reference: _____

11. Can Type NM cable be used for a 120/240-volt branch circuit as temporary wiring in a building under construction where the cable is supported on insulators at intervals of not more than 10 feet (3 m)?

Answer: _____

_____

_____

_____

Reference: _____

12. Which chapter of the *NEC®* is independent of all other chapters?

Answer: _____

_____

_____

_____

Reference: _____

13. Is the *NEC®* considered a training manual?

Answer: _____

_____

_____

_____

Reference: _____

14. Are all mining facilities exempt from the *NEC®*?

Answer: _____

_____

_____

_____

Reference: _____

15. A comment often made by those involved in electrical design or installation states that the *NEC*® is a minimum standard. Where in the *NEC*® does it state that it is the true minimum permitted for electrical installations?

Answer: _____

_____

_____

_____

Reference: _____

16. As the inspector was making an inspection and looked at the size of overcurrent device and conductors supplying a motor, the inspector asked, "Is this motor circuit rated for continuous duty?" What is a continuous load?

Answer: _____

_____

_____

_____

Reference: _____

17. When you arrive in a nearby city to make an installation, the electrical inspector informs you that the city has not adopted the last two editions of the *NEC*®. Which edition would that city be enforcing?

Answer: _____

_____

_____

_____

Reference: _____

18. In wiring a small residence, the owners inform you that they wanted a doorbell mounted on the front and back of the house. What class wiring would this small 24-volt doorbell circuit be wired in and what article governs that wiring method?

Answer: _____

_____

_____

_____

Reference: _____

19. We have been asked to bid a nursing home where patients will have varied degrees of mobility. Some will come and go freely, cook their own meals, and do their own housekeeping; others may require meals to be prepared and minor medical attention, such as someone ensuring that they take their medicine regularly. Other patients may be bedridden and require oxygen or a doctor's care. Which article of the *NEC*® covers a nursing home facility of this type?

Answer: _____

_____

_____

_____

Reference: _____

20. In establishing a grounding electrode on a new installation, the owner says that although an underground metal water pipe exists, the owner prefers that it not be used and that a standard 8-foot (2.44 m) ground rod be used instead. Which section of the *NEC*® governs grounding electrodes?

Answer: _____

_____

_____

_____

Reference: _____

Chapter Three

BASIC ELECTRICAL MATHEMATICS REVIEW

In this chapter, we refresh our knowledge of the basic mathematical calculations needed to perform the simple day-to-day mathematical tasks required in our work as electrical contractors or electricians. Many tasks reviewed in this chapter may appear simple. However, if you have difficulty in solving any of these problems, then it may be necessary for you to obtain a more complete study guide on basic electrical mathematics. This chapter is not only a mathematical teaching aid but also a refresher to see that your mathematical skills are sufficient to solve the problems that you will encounter in most electrical examinations.

CALCULATOR MATH

Many of us were taught and still believe that mathematical study should be done manually. But in today's world, with the introduction of handheld calculators into the industrialized world in the past 35 years, it would be somewhat silly to depend on all hand calculations. Most testing jurisdictions permit the use of handheld, silent, battery-powered, nonprogrammable calculators to use during the test. Therefore, it is advantageous to work the problems in this chapter using the same calculator that you will take to the examination. Many calculators are on the market, priced from advertising giveaways to complex scientific notation calculators. However, for your purposes, any quality handheld calculator will do. It should have the standard engineering functions, but it is not necessary to have one with scientific capabilities. The button arrangement should be good quality. It should be battery-powered, not solar-powered, because the room lighting might be inadequate for solar power. You should have a memory function and most modern mathematical functions, including the basic keys, input, error correction, combining operations, calculator hierarchy (calculations with a constant), roots, powers, reciprocal, factorial, percents, natural logarithm and natural antilogarithm, trigonometric functions, and error indication, accuracy and rounding, good memory usage for storing, recalling, and memory exchange, conversion factors from English to metric, and temperature conversions. A calculator of this quality is available in most discount stores for under $20.00. Get the calculator that you plan to use and familiarize yourself with it so that you are familiar with its operations before the examination.

The Basic Keys

+	−	×	÷	=

Example: If you have $300.00 in your checking account, and you write checks for $6.00, $35.00, $53.00, and $125.25 and make a deposit for $200.00, what is your balance?

$300 − 6 − 35 − 53 − 125.25 + 200 = $280.75 balance

Example: If 250 feet of Type NM Cable costs $31.75, how much does 190 feet cost?

$31.75 ÷ 250 = .127 × 190 = $24.13 cost

Caution: Do not rely on programmable calculators to verify calculations or to prepare for the examination. They are unacceptable and are not permitted in most testing organizations.

Figure 3–1 An example of a handheld calculator suitable for quick field calculation. (*Courtesy of Calculated Industries, Inc.*)

Although programmable calculators are not permitted in most testing organizations, there are good programmable handheld calculators on the market as shown in Figure 3–1 and Figure 3–2. Although these are unacceptable in an examination, they can provide a convenient tool for the contractor or electrician or those studying to improve their skills in the electrical industry.

WORKING WITH FRACTIONS

Assuming that your basic addition, subtraction, multiplication, and division skills are still sharp, we begin with a brief review of fractions. A fraction is a part of a number. For instance, if we have a pie, we have one whole pie. If we cut that pie into four pieces, then each piece is one fourth of a pie. The four pieces together equal four fourths, or one. If we cut that whole pie into five pieces, then each piece would be one fifth of the pie. The five pieces together would be one pie.

If we have two pieces of a pie that has been cut into four pieces, we would have two fourths. This fraction can be expressed in two different ways—by ²⁄₄ or ½. To reduce a fraction is to change it into another equal fraction. To do this, you divide the numerator (top number) and the denominator (bottom number) by the same number. For instance, to reduce ²⁄₆, divide the 2 and the 6 by 2. Then the numerator would be 2 divided by 2, or 1. The denominator would be 6 divided by 2, which equals 3. Therefore, the reduced fraction is ⅓; ²⁄₆ = ⅓.

Example: $\dfrac{2}{6} = \dfrac{2/2}{6/2} = \dfrac{1}{3}$

To reduce larger fractions, such as ²⁴⁄₉₆, to the lowest form, we can divide both the 24 and the 96 by 24 (to get ¼) or we can divide the 24 and the 96 by 8 (which gives ³⁄₁₂) and then divide the 3 and the 12 by 3, which gives ¼, the answer.

Example: $\dfrac{24}{96} = \dfrac{24/24}{96/24} = \dfrac{1}{4}$

Reducing Mixed Numbers to Improper Fractions

In a proper fraction, the numerator is always smaller than the denominator, such as ⅓, ⅙, ¹⁄₁₂. Mixed numbers are made of two numbers, a whole number and a fraction, for example: 2½, 5½, 3³⁄₁₆, numbers that we use every day in our work. To change these to improper fractions, the numerator will always be larger than the denominator, such as ¹²⁄₃ or ⁹⁄₆ or ²⁴⁄₁₀. For example, to change 2½ to an improper fraction, the 2 will be a certain number of halves (2 = ⁴⁄₂), which we then add to the fraction we are given (½). Thus, 2½ = ⁴⁄₂ + ½ = ⁵⁄₂.

Example: $2\dfrac{1}{2} = \dfrac{2 \times 2 + 1}{2} = \dfrac{5}{2}$

To change a mixed number to an improper fraction, multiply the denominator of the fraction by the whole number and add the numerator of the fraction. Place this answer over the denominator to make an improper fraction.

Example: $5\dfrac{1}{2} = \dfrac{2 \times 5 + 1}{2} = \dfrac{11}{2}$

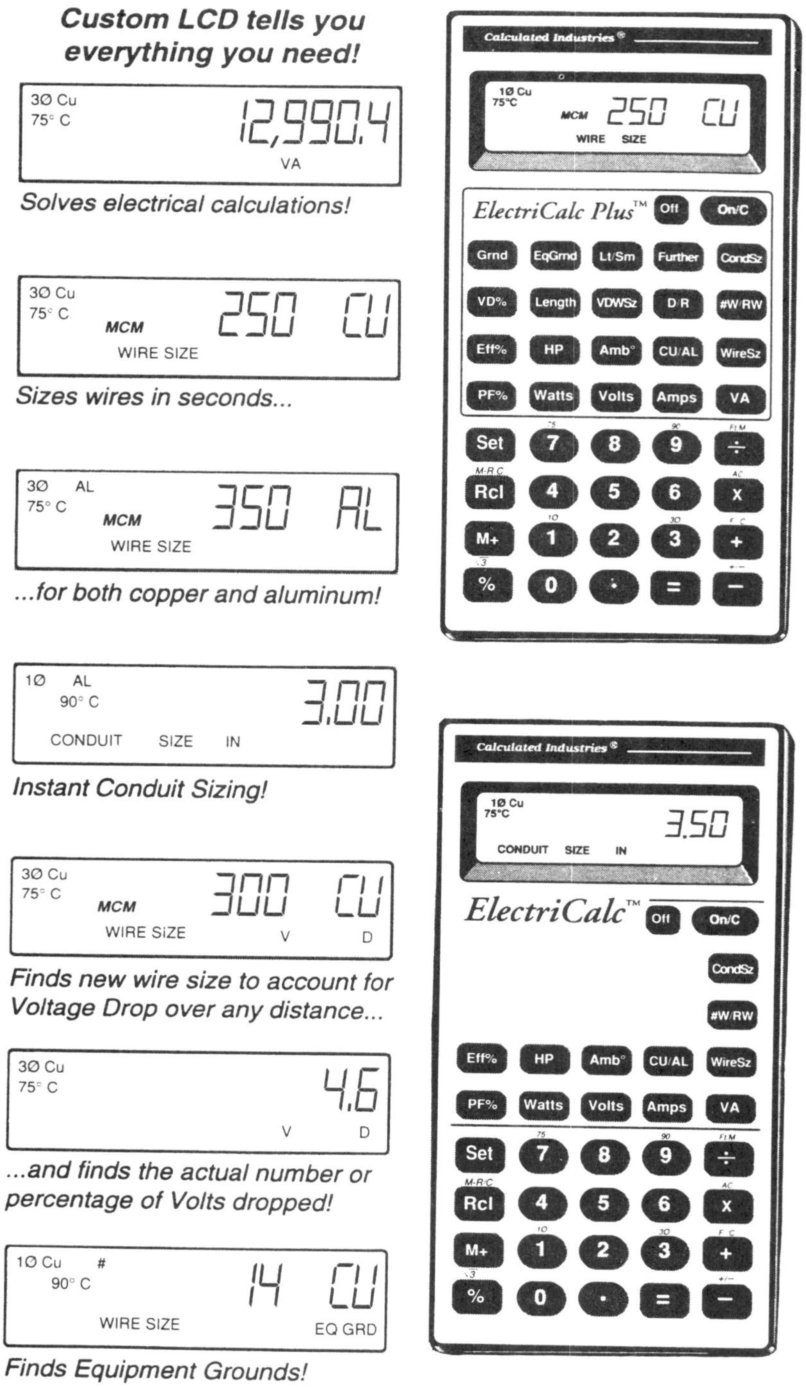
Custom LCD tells you everything you need!
3Ø Cu 75° C 12,990.4 VA
Solves electrical calculations!
3Ø Cu 75° C MCM 250 CU WIRE SIZE
Sizes wires in seconds...
3Ø AL 75° C MCM 350 AL WIRE SIZE
...for both copper and aluminum!
1Ø AL 90° C 3.00 CONDUIT SIZE IN
Instant Conduit Sizing!
3Ø Cu 75° C MCM 300 CU WIRE SIZE V D
Finds new wire size to account for Voltage Drop over any distance...
3Ø Cu 75° C 4.6 V D
...and finds the actual number or percentage of Volts dropped!
1Ø Cu # 90° C 14 CU WIRE SIZE EQ GRD
Finds Equipment Grounds!
Calculated Industries
ElectriCalc Plus™
ElectriCalc™

Figure 3–2 Examples of some of the types of multifunction tasks that can be quickly solved. (*Courtesy of Calculated Industries, Inc.)*

Changing Improper Fractions to Mixed Numbers

To change an improper fraction, such as 13⁄3, to a mixed number, divide the numerator by the denominator (13 divided by 3). Any remainder is placed over the denominator (13 divided by 3 = 4 with a remainder of 1). The resulting whole number and proper fraction form a mixed number. Thus, $\frac{13}{3} = 4 + \frac{1}{3} = 4\frac{1}{3}$.

Example: $\dfrac{13}{3} = 4 + \dfrac{1}{3} = 4\dfrac{1}{3}$

Multiplication of Fractions

To multiply fractions, place the multiplication of the numerators over the multiplication of the denominators and reduce to lowest terms. For example: ½ × ¾. Answer: ½ × ¾ means that 1 × 3 = 3 and 2 × 4 = 8; thus, the answer is ⅜. Example: ½ × 5. Answer: ½ × 5 means that 1 × 5 = 5 and 2 × 1 = 2, giving an improper fraction of ⁵⁄₂, which then has to be reduced to 2½.

Examples: $\dfrac{1}{2} \times \dfrac{3}{4} = \dfrac{1 \times 3}{2 \times 4} = \dfrac{3}{8}$

$\dfrac{1}{2} \times 5 = \dfrac{1 \times 5}{2 \times 1} = \dfrac{5}{2} = 2\dfrac{1}{2}$

To cancel the numerator and denominator means to divide both numerator and denominator by the same number. As an example, when multiplying ⅜ × ⁴⁄₉, notice that the 3 on top and the 9 on the bottom can both be divided by 3. Cross out the 3 and write a 1 over it. Cross out the 9 and write a 3 under it. Also, the 4 on the top and the 8 on the bottom can both be divided by 4. Cross out the 4 and write a 1 over it. Cross out the 8 and write a 2 under it. The solution then is 3 reduced to 1 × 4 reduced to 1 (1 × 1) over 8 reduced to 2 × 9 reduced to 3 (2 × 3), thus giving ⅙.

Example:

$$\dfrac{3}{8} \times \dfrac{4}{9} = \dfrac{\overset{1}{\cancel{3}} \times \overset{1}{\cancel{4}}}{\underset{2}{\cancel{8}} \times \underset{3}{\cancel{9}}} = \dfrac{1}{6}$$

WORKING WITH DECIMALS

A decimal is a fraction in which the denominator is not written. The denominator is 1. Many answers to electrical problems will be fractions, such as ¼ of an ampere, ½ of an ampere, ½ of a volt, ⅓ of a volt, and the like. These will be correct mathematical answers, but they will be worthless to an electrician. The electrical measuring instruments give values expressed in decimals, not in fractions. In addition, the manufacturers of components give the values of parts in terms of decimals, such as 3.4 amperes, and the like, on the nameplate of a motor. Suppose we worked out a problem and found that the current in the circuit should be ⅛ of an ampere, and using the ampmeter, we test the circuit and find that .125 ampere flows. Is our circuit correct? How will we know? How can we compare

⅛ to .125? The easiest way is to change the fraction ⅛ to its equivalent decimal and then compare the decimals. To change a fraction into a decimal, divide the numerator by the denominator, (1 divided by 8 = .125).

Example: $\dfrac{1}{8} = \dfrac{1}{8.000} = .125$

Converting Fractions to Decimals
Divide the numerator (top number) by the denominator (bottom number).

½ = 1 ÷ 2 = 0.5
¾ = 3 ÷ 4 = 0.75
⅞ = 7 ÷ 8 = 0.875

$9\tfrac{3}{8} = \dfrac{9 \times 8 + 3}{8} = \dfrac{75}{8} = 75 \div 8 = 9.375$

Converting Percentages to Decimals
Move decimal two places to the left.

100% = 1.00
50% = 0.50
37½% = 0.375

WORKING WITH PERCENTAGES

To apply the rules of the *NEC®*, you will need to know how to work with percentages—for example, **430.110 (A) General.** *The disconnecting means for motor circuits rated 600 volts, nominal, or less shall have an ampere rating of at least 115 percent of the full-load current rating of the motor.* [Reprinted with permission from NFPA 70-2005.]

Question: A 480-volt motor has a full-load current of 28 amperes. What size disconnecting means is required?

    A. 20 amperes
    B. 25 amperes
    C. 30 amperes
    D. 40 amperes

Because the motor load is 28 amperes, you can quickly eliminate answers A and B.

To solve: 28 × 115% = 28 × 1.15 = 32.2 = 40 amperes disconnecting means required.

Question: A custom home (single-family dwelling) has branch-circuit general illumination load of 200,000 VA (volt-amperes). What demand factor must be applied to this general illumination load?

    A. 200,000
    B. 43,950
    C. 105,000
    D. 63,950

Answer: D

To solve:

First 3000 at 100% = 3000 × 1	= 3000 VA
3001 to 120,000 = 117,000 at 35% = 117,000 × .35	= 40,950 VA
Over 120,000 = 80,000 at 25% = 80,000 × .25	= 20,000 VA
Total 200,000 VA lighting load demand	63,950 VA

220.42 General Lighting. *The demand factors specified in Table 220.42 shall apply to that portion of the total branch-circuit load calculated for general illumination. They shall not be applied in determining the number of branch-circuits for general illumination.* [Reprinted with permission from NFPA 70-2005.]

WORKING WITH SQUARE ROOT

Roots are the opposite of powers. A square root is the opposite of the number to the second power. The sign of the square root is $\sqrt{}$. To find a square root, find a number that when multiplied by itself is the number inside the square root sign. For example, to find the value of the square root of 36, ask yourself what number times itself is 36. The answer is 6, so 6 is the square root of 36.

If you can find the square root of a number with the method that uses averages, suppose that you did not know that the square root of 144 is 12. When you divide the number by its square root, the answer is the square root. If you cannot find this answer, then guess as close as possible. A good guess for 144 would be 10, because 10 × 10 = 100, which is close to 144. Divide 144 by 10 and the answer is 14 + a remainder. Average the guess, 10, and the answer to the division problem, 14. 10 + 14 = 24, $^{24}/_{2}$ = 12, which is the correct answer.

Follow these steps to find the square root of the larger number: Guess the answer; divide the guess into the large number; average the guess and the answer to

Table 220.42 Lighting Load Demand Factors

Type of Occupancy	Portion of Lighting Load to Which Demand Factor Applies (Volt-Amperes)	Demand Factor (Percent)
Dwelling units	First 3000 or less at	100
	From 3001 to 120,000 at	35
	Remainder over 120,000 at	25
Hospitals*	First 50,000 or less at	40
	Remainder over 50,000 at	20
Hotels and motels, including apartment houses without provision for cooking by tenants*	First 20,000 or less at	50
	From 20,001 to 100,000 at	40
	Remainder over 100,000 at	30
Warehouses (storage)	First 12,500 or less at	100
	Remainder over 12,500 at	50
All others	Total volt-amperes	100

*The demand factors of this table shall not apply to the calculated load of feeders or services supplying areas in hospitals, hotels, and motels where the entire lighting is likely to be used at one time, as in operating rooms, ballrooms, or dining rooms.

Reprinted with permission from NFPA 70-2005.

the division problem; and check. For example, find the value of the square root of 1024.

Step 1. Guess: in the list of square roots, 30 ′ 30 = 900. This is too small, but it is easy to divide by.

Step 2. Divide 1024 by 30. The answer is 34, with a remainder.

Step 3. Find the average of 30 and 34. 30 + 34 = 64; $^{64}/_{2}$ = 32.

Step 4. Check: multiply 32 × 32. The answer is 1024. Thus, 32 is the square root of 1024. When you use this method to find square roots, always guess a number that ends in 0. It is easier and faster to divide by these numbers. If the average is not the square root of the number, use the average as a new guess and try again.

Finding the Reciprocal of a Number

There are instances in the *NEC®* where it is easier to find the reciprocal of a number and then make the calculation, rather than use the number in the calculation. For example, the reciprocal of 125% is 80%, and the reciprocal of 1.73 is .578, which may be rounded off to .58 for quick calculations.

To find a reciprocal of a number, divide the number into 1.

$$1 \div 1.73 = .5780 \text{ or } .58$$

If you are finding the reciprocal of a percentage, you must first convert the percent to a decimal.

$$125\% = 1 \div 1.25 = 0.8 \text{ or } 80\%$$

(E) Single Non–motor-Operated Appliance. *If the branch-circuit supplies a single non–motor-operated appliance, the rating of overcurrent protection shall :*
(1) Not exceed that marked on the appliance;
(2) Not exceed 20 amperes if the overcurrent protection rating is not marked and the appliance is rated 13.3 amperes or less; or
(3) Not exceed 150 percent of the appliance rated current if the overcurrent protection rating is not marked and the appliance is rated over 13.3 amperes. Where 150 percent of the appliance rating does not correspond to a standard overcurrent device ampere rating, the next higher standard rating shall be permitted. [Reprinted with permission from NFPA 70-2005.]

Question: A quick recovery electrical water heater is rated at 4500 watts 240 volts. Is a 30-ampere circuit breaker permitted to protect this appliance?

Answer: Yes, 4500 ÷ 240 = 18.75 × 150% = 28.1 = the next higher standard circuit breaker rating is 30 amperes.

You may first think a 20-ampere circuit breaker would be acceptable for a 18.75-ampere load. However, 422.10(A) requires the branch-circuit rating to be at least 125 pecent of the marked rating. A 20-ampere circuit breaker is permitted to protect the reciprocal of 125 percent or 80 percent of the circuit breaker rating. Thus 80% × 20 = 16 amperes.

A 25-ampere circuit breaker would be acceptable. However, a 25-ampere circuit breaker is not a common size and may not be readily available.

POWERS

A power is a product of a number multiplied by itself one or more times. For an example, what is the value of 3^2? The exponent is 2; write 3 two times, or $3^2 = 3 \times 3$; $3^2 = 9$. What is the value of 5^3? The exponent is 3; write 5 three times; in other words, $5^3 = 5 \times 5 \times 5 = 125$. These powers are terms often used in the electrical field and should be known.

OHM'S LAW

E = voltage; I = current (amperes); R = resistance

$$E = IR; \; I = \frac{E}{R}; \; R = \frac{E}{I}$$

As an example, all numbers are relative. For example, if E = 12, I = 3, and R = 4, then $3 \times 4 = 12$, $\frac{12}{3} = 4$, and $\frac{12}{4} = 3$. All numbers are relative. Remember, in any electrical circuit, the voltage force is the current through the conductor against its resistance. The resistance tries to stop the current from flowing. The current that flows in the circuit depends on the voltage and the resistance. The relationship among these three quantities is described by Ohm's law. Ohm's law applies to an entire circuit or any component part of a circuit.

Voltage Drop Calculations

We have discussed the methods for finding the resistance by using Ohm's law. Now we can find the voltage drop for circuit loads using this same formula.

$$E = I \times R$$

Voltage Drop Formula

Single-phase voltage drop = amperes × length × resistance × (2)

Voltage drop % = voltage drop × 100/volts

Three-phase voltage drop—calculate as a single-phase circuit and multiply resultant by .866.

$$V_{(D)} = I \times 2L \times R$$

I = Amperes
L = Length of circuit
R = Resistance values from *NEC®* Chapter 9, Table 8
Resistance = R × [1 + a (temperature − 75)]
a = 0.00323 for copper, 0.000330 for aluminum
Temperature = ambient temperature

Voltage drop can be calculated with the known elements of the power circuit. The resistance of conductors commonly used for electrical power conductors can be found in *NEC®* Chapter 9, Table 8.

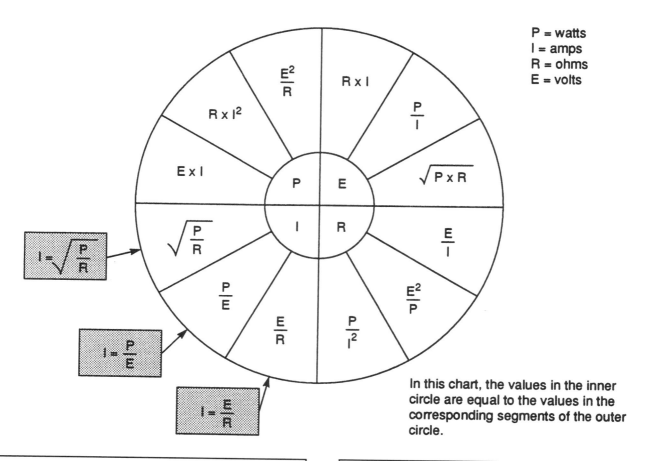

P = watts
I = amps
R = ohms
E = volts

$I = \sqrt{\dfrac{P}{R}}$

$I = \dfrac{P}{E}$

$I = \dfrac{E}{R}$

In this chart, the values in the inner circle are equal to the values in the corresponding segments of the outer circle.

Example: What is the voltage drop for a 120-volt, 20-ampere, single-phase circuit supplying a 12-ampere load located 125 feet (37.5 m) away from the source of supply? The circuit conductors have a combined resistance of 0.35 ohm.

A. 21 volts
B. 42 volts
C. 2.1 volts
D. 4.2 volts

Answer: D

What is the question? What is the voltage drop? **Ohm's law works!**

$E = IR$ $I = 12$ amperes $R = 0.35$ ohm
$E = 12 \times 0.35$
$E = 4.2$ volts

Note that this is a 3½ percent voltage drop. 210.19 FPN No. 4 recommends that branch-circuits do not exceed a 3 percent drop. For maximum efficiency, the circuit conductors should be increased in size.

NEC® 2005 210.19(A)(1) FPN No. 4: Conductors for branch circuits as defined in Article 100, sized to prevent a voltage drop exceeding 3 percent at the

farthest outlet of power, heating, and lighting loads, or combinations of such loads, and where the maximum total voltage drop on both feeders and branch circuits to the farthest outlet does not exceed 5 percent, provides reasonable efficiency of operation. See FPN No. 2 of 215.2(A)(3) for voltage drop on feeder conductors. *

The *NEC®* references prescribe voltage drop values or percentages for feeder conductors and branch-circuit conductors; however, it should be noted these appear in a fine print note (FPN) and are only suggested and not mandatory requirements. The *NEC®* 215.2(A)(3) FPN 2 recommends voltage drop does not exceed 3 percent of the farthest outlet of power, heating, or lighting loads or combinations of such loads, and the maximum voltage drop for both feeders and branch circuits to the farthest outlet is not to exceed 5 percent and will provide reasonable efficiency of operation.

Example: A dc circuit has 2 amperes of current flowing in it. A voltmeter reads 10 volts line to line. How much resistance is in the circuit? Answer: 5 ohms.

* Reprinted with permission from NFPA 70-2005.

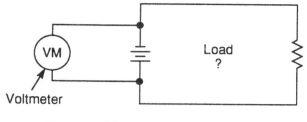

$$R = \frac{E}{I} \quad R = \frac{10}{2} \quad R = 5 \text{ ohms}$$

Example: If the voltage is 100 volts and the resistance is 25 ohms, what amount of current will flow in the circuit? Answer: 4 amperes.

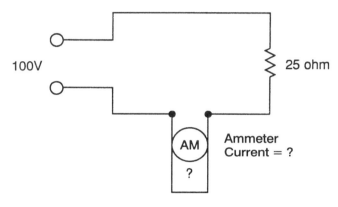

$$I = \frac{E}{R} \quad I = \frac{100}{25} \quad I = 4 \text{ amperes}$$

Example: If the potential across a circuit is 120 volts and the current is 6 amperes, what is the resistance? Answer: 20 ohms.

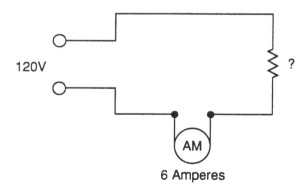

$$R = \frac{E}{I} \quad R = \frac{120}{6} \quad R = 20 \text{ ohms}$$

A series circuit may be defined as a circuit in which the resistive elements are connected in a continuous run (end to end). It is evident that because the cir-

cuit has no branches, the same current flows in each resistance. The total potential across the circuit equals the sum of potential drops across each resistance.

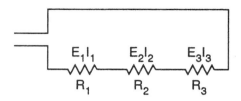

Example:
$$E_1 = IR_1$$
$$E_2 = IR_2$$
$$E_3 = IR_3$$
$$E = E_1 + E_2 + E_3$$
$$R = R_1 + R_2 + R_3$$

The total potential of the circuit is:

$$E = 1R_1 + 1R_2 + 1R_3 = I(R_1 + R_2 + R_3)$$

$$I = \frac{E}{R_1 + R_2 + R_3}$$

$R_1 = 10$ ohms, $R_2 = 5$ ohms, and $R_3 = 15$ ohms.

What amount of voltage must flow to force 0.5 amperes through the circuit?

$$R_T = 5 + 10 + 15 = 30 \text{ ohms}$$

Hence, $E_T = 0.5 \times 30 = 15$ volts

What is the voltage drop across each resistance?

$$E_1 = 0.5 \times 10 = 5.0 \text{ volts}$$
$$E_2 = 0.5 \times 5 = 2.5 \text{ volts}$$
$$E_3 = 0.5 \times 15 = \underline{7.5 \text{ volts}}$$
$$E_T = 15 \text{ volts}$$

When the resistance and current are known, what is the formula to determine the voltage?

The current is 2 amperes, the resistance is 15 ohms, 10 ohms, and 30 ohms.

$$E = IR \qquad E = 2(15 + 10 + 30)$$
$$E = 2 \times 55 \qquad E = 110 \text{ volts}$$

In the industry today most circuits are connected in parallel. In a parallel circuit the voltage is the same across each resistance (load).

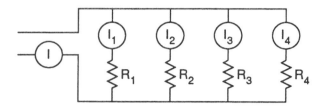

The total current is equal to the sum of all the currents.

$$E = I_1R_1 = I_2R_2 = I_3R_3 = I_4R_4$$

and

$$I = I_1 + I_2 + I_3$$

Individual Resistances =

$$I_1 = \frac{E}{R_1} \quad I_2 = \frac{E}{R_2} \quad I_3 = \frac{E}{R_3} \quad I_4 = \frac{E}{R_4}$$

Hence: $I_1 = \frac{E}{R_1} + I_2 = \frac{E}{R_2} + I_3 = \frac{E}{R_3} + I_4 = \frac{E}{R_4}$

or $I = E\left(\frac{1}{R_1} + \frac{1}{R_2} + \frac{1}{R_3} + \frac{1}{R_4}\right)$

and several resistances in parallel is

$$\frac{1}{R} = \frac{1}{R_1} + \frac{1}{R_2} + \frac{1}{R_3} + \frac{1}{R_4}$$

The sum of the resistances in parallel always will be smaller than the smallest resistance in the circuit where only two resistances are in parallel.

$$R = \frac{R_1 \times R_2}{R_1 + R_2} \text{ ohms}$$

Power Factor

Power factor is a phase displacement of current and voltage in an ac circuit. The cosine of the phase angle of displacement is the power factor. The cosine is multiplied by 100 and is expressed as a percentage. A cosine of 90° is 0; therefore, the power factor is 0 percent. If the angle of displacement were 60°, the cosine of which is .5, the power factor would be 50 percent. This is true whether the current leads or lags the voltage. Power is expressed in dc circuits and ac circuits that are purely resistive in nature. Where these circuits contain only resistance, P (watts) = E × I. In ac circuits that contain inductive or capacity reactances, VA (volt-ampere) = E × I. In a 60-cycle ac circuit, if the voltage is 120 volts, the current is 12 amperes, and the current lags the voltage by 60°, find (a) the power factor, (b) the power and voltage-amperes (VA), and (c) the power in watts. The cosine of 60 is .5; therefore, the power factor is 50 percent. 120 × 12 = 1440 volt-amperes, which is called the apparent power. 120 × 12 × .5 = 720 watts, which is called the true power. Power factor is important. There is an apparent power of 1440 VA and a true power of 720 watts. There also are 12 amperes of line current and 6 amperes of in-phase or effective current. This means that all equipment from the source of supply to the power consumption device must be capable of handling a current of 12 amperes, when actually the device is only using the current of 6 amperes. A 50 percent power factor was used intentionally to make the results more pronounced. The I²R loss is based on the 12-ampere current, while only 6 amperes are effective. Power factor can be measured by a combined use of a voltmeter, ammeter, and watt-meter, or by the use of a power factor meter. When using the three meters, volt-ampere meter and watt-meter, all connected properly in the circuit, the readings of the three meters are taken simultaneously under the same load conditions and calculated as follows: power factor = true power (watts)/apparent power (which is volt-amperes). Power factor = W/EI.

1000 watts = 1 KW
1000 watts operating for 1 hour = 1 KWH used
3413 BTU = 1 KW
3.413 BTU = 1 watt
Fahrenheit = (9/5 × C) + 32
Celsius = 5/9 × (F − 32)
P. F. = KW / KVA
KVA = KW / P. F.

QUESTION REVIEW

Directions: Reduce the following fractions to their lowest terms.

1. $\frac{3}{6}$ = _____

2. $\frac{4}{16}$ = _____

3. $\frac{4}{40}$ = _____

4. $\frac{40}{100}$ = _____

5. $\frac{6}{8}$ = _____

6. $\frac{15}{20}$ = _____

7. $\frac{18}{36}$ = _____

8. $\frac{9}{15}$ = _____

9. $\frac{30}{50}$ = _____

10. $\frac{4}{20}$ = _____

11. $\frac{5}{20}$ = _____

12. $\frac{6}{24}$ = _____

Directions: Change the following mixed numbers to improper fractions.

1. $1\frac{1}{2}$ = _____

2. $3\frac{1}{4}$ = _____

3. $5\frac{3}{8}$ = _____

4. $4\frac{1}{4}$ = _____

5. $7\frac{1}{8}$ = _____

6. $2\frac{1}{6}$ = _____

7. $5\frac{3}{4}$ = _____

8. $8\frac{1}{3}$ = _____

9. $6\frac{4}{7}$ = _____

10. $3\frac{5}{8}$ = _____

Directions: Change the following improper fractions to mixed numbers.

1. $\frac{9}{2}$ = _____

2. $\frac{12}{5}$ = _____

3. $\frac{64}{5}$ = _____

4. $\frac{26}{8}$ = _____

5. $\frac{29}{3}$ = _____

6. $\frac{31}{2}$ = _____

7. $\frac{5}{3}$ = _____

8. $\frac{13}{3}$ = _____

Directions: Multiply the following whole numbers and fractions. Give the answer in proper reduced form.

1. $8 \times \frac{1}{2}$ = _____

2. $\frac{1}{2} \times \frac{3}{4}$ = _____

3. $\frac{1}{5} \times \frac{1}{8}$ = _____

4. $9 \times \frac{1}{2}$ = _____

5. $\frac{3}{4} \times 7$ = _____

6. $\frac{1}{2} \times 5$ = _____

7. $\frac{3}{4} \times 12$ = _____

8. $\frac{1}{7} \times 15$ = _____

Directions: Convert the following fractions to decimal equivalents.

1. $\frac{1}{4}$ = _____

2. $\frac{5}{8}$ = _____

3. $\frac{3}{4}$ = _____

4. $\frac{13}{15}$ = _____

5. $\frac{3}{8}$ = _____

Directions: Resolve to whole numbers.

1. 2^4 = _____

2. 9^2 = _____

3. 50^2 = _____

4. 1^5 = _____

5. 10^3 = _____

Directions: Find the square root of the following numbers.

1. Square root of 25 = _____

2. Square root of 81 = _____

3. Square root of 49 = _____

4. Square root of 3136 = _____

5. Square root of 10,201 = _____

6. Square root of 841 = _____

7. Square root of 6064 = _____

Directions: Solve the following problems.

1. A 20-ampere load is fed with two conductors that have a *combined* resistance of 0.3 ohm. If the source voltage is 120 volts dc, the calculated voltage drop on this circuit is _____ percent.

   Answer: _____

   _____

   _____

   _____

   Reference: _____

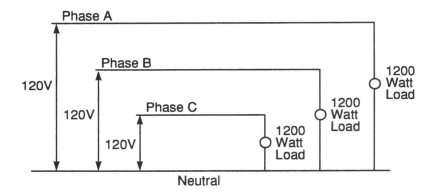

Phase A

Phase B

Phase C

120V

120V

120V

1200 Watt Load

1200 Watt Load

1200 Watt Load

Neutral

2. Refer to the above drawing.

    I(LINE A) = I(LINE B) = I(LINE C) = 10 amperes.

The power factor in the circuit in the figure above is _____ .

Answer: _____

_____

_____

_____

Reference: _____

3. A 120-volt branch circuit has only six 100-watt, 120-volt incandescent luminaires (fixtures) connected to it. The current in the home run of this circuit is _____ amperes.

Answer: _____

_____

_____

_____

Reference: _____

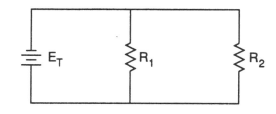

E_T R_1 R_2

4. Refer to the above drawing.

    With a resistance at R_1 of 4 ohms and a resistance at R_2 of 2 ohms, the total resistance for the above circuit is _____ ohms.

Answer: _____

_____

_____

_____

Reference: _____

5. A 1000-watt, 120-volt lamp uses electrical energy at the same rate as a/an _____ ohm resistor.

Answer: _____

_____

_____

_____

Reference: _____

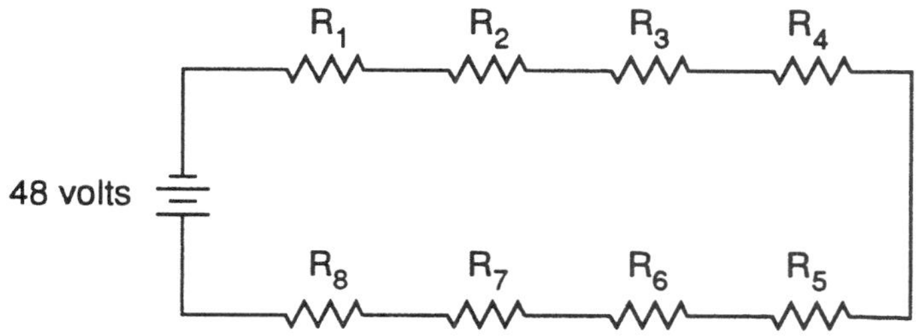

6. In the above diagram, eight equal resistors are connected in series to a 48-volt source. The voltage drop across each resistor is _____ volts.

Answer: _____

_____

_____

_____

Reference: _____

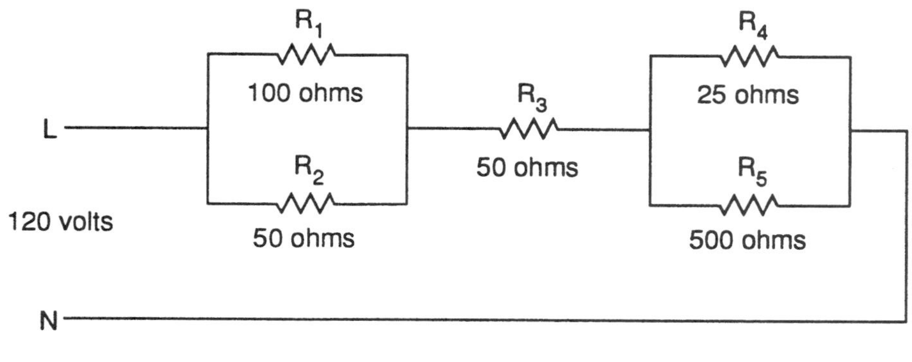

The above diagram applies to Questions 7 through 14.

7. What is the resistance of R_1 and R_2?

Answer: _____

_____

_____

_____

Reference: _____

8. What is the resistance of $R_3 + R_1$ and R_2?

   Answer: _____

   _____

   _____

   _____

   Reference: _____

9. What is the resistance of R_4 and R_5?

   Answer: _____

   _____

   _____

   _____

   Reference: _____

10. What is the total resistance in the circuit?

    Answer: _____

    _____

    _____

    _____

    Reference: _____

11. What is the voltage drop at R_1 and R_2?

    Answer: _____

    _____

    _____

    _____

    Reference: _____

12. What is the voltage drop at R3?

    Answer: _____

    _____

    _____

    _____

    Reference: _____

13. What is the voltage drop at R_4 and R_5?

    Answer: _____

    _____

    _____

    _____

    Reference: _____

14. What is the total current in this circuit?

    Answer: _____

    _____

    _____

    _____

    Reference: _____

PRACTICE EXAM

1. 0.875 is the equivalent of which of the following?
   A. 9/32
   B. 15/16
   C. 7/8
   D. 5/8

2. How much current will a 240-volt circuit with 20-ohm load draw?
   A. 24 amperes
   B. 2.4 amperes
   C. 1.2 amperes
   D. 12 amperes

3. 4.75 is the equivalent of which of the following improper fractions?
   A. 23/4
   B. 17/5
   C. 15/3
   D. 38/8

4. A 10 kW load is increased by 12 percent every summer. What is the VA in the summer?
   A. 11.2
   B. 1120
   C. 11,200
   D. 112,000

5. What is the circumference of a circle with a 1-inch diameter?
   A. 1 inch
   B. 1.2 inches
   C. 3.14 inches
   D. 2½ inches

6. What is the radius of a circle with a 1-inch diameter?
   A. ½ inch
   B. 2½ inches
   C. 1 inch
   D. 1½ inches

7. What is the resistance of a 20-ampere, 120-volt circuit?
   A. 5 ohms
   B. .5 ohm
   C. .6 ohm
   D. 6 ohms

8. One meter equals how many feet?
   A. 32 inches
   B. 3.28 feet
   C. 2.88 feet
   D. 1½ feet

9. One meter equals how many millimeters?
   A. 100
   B. 10
   C. 1000
   D. 10,000

10. According to the *NEC*®, 10 feet converted to metric (soft conversion) equals
    A. 3 m.
    B. 3.05 m.
    C. 30 m.
    D. 300 m.

11. According to the *NEC*®, 3 feet converted to metric (hard conversion) equals
    A. .3 m.
    B. 300 mm.
    C. 900 mm.
    D. 9 m.

12. How many amperes does a 3-phase, 480-volt, 24-kW load draw? Round off to a whole number.
    A. 29
    B. 50
    C. 290
    D. 5

13. ¾ can be also shown as what percent?
    A. .75%
    B. 75%
    C. ¾%
    D. none of the above

14. Unless specifically permitted in the *NEC*®, 12 AWG conductors, Type THWN overcurrent protection shall not exceed
    A. 16 amperes.
    B. 25 amperes.
    C. 20 amperes.
    D. 30 amperes.

15. The reciprocal of 125 percent is _____.
    A. 0.8
    B. .58
    C. 1.73
    D. 1.25

Chapter Four

OBJECTIVES

Studying this chapter along with the 2005 *NEC*® will give the reader a basic introduction to the *Code* and an understanding of how to start using the 2005 *NEC*® efficiently.

After completing this chapter, you should know:

- The unique style the *NEC*® is written in
- The types of installations covered and exempted
- The application of Chapters 1–4 generally
- The application of Chapters 1–5 to supplement and modify the general requirements
- The application of Chapter 8 independently
- The application of the tables in Chapter 9
- The Annex for information and examples
- The importance of thoroughly understanding contents of
  1. Article 90—Introduction
  2. Article 100, definitions of words and terms unique to the *NEC*®
  3. Article 110, general requirements that apply throughout
  4. Article 300, general wiring applications that apply to all wiring methods

INTRODUCTION TO THE *NATIONAL ELECTRICAL CODE*® (*NEC*®)

The *National Electrical Code*® is a reasonable document, developed and written in a reasonable fashion. It is a design manual, but it is not a design specification. It is not the minimum, but there are many conveniences that we take for granted that are not required by this document. There are also many safety considerations that are not required by this text, such as smoke detectors and other safety and detection devices that are being developed every day. The first page in Article 90, 90.1(C), says it is not an instruction manual for untrained persons. You are going to find that that statement explains much of the confusion about this document. Most engineering schools, vocational schools, and apprentice programs do not have training

programs on the proper use of this document as an installation and design standard. Open your 2005 *NEC*® and look up *NEC*® 90.1(C). You will need the 2005 *National Electrical Code*® (*NEC*®) throughout your study of this manual. There are several *NEC*® handbooks available. The NFPA *National Electrical Code*® *Handbook* contains the exact *NEC*® text and commentary. If you are using that text, remember that the commentary is only the author's opinion and may not be totally correct. Therefore, when studying in a classroom or for an examination, the *NEC*® text will be that from which questions are developed.

KNOW YOUR CODE BOOK

To use the *NEC*®, we must first have a thorough understanding of:

- the Table of Contents
- the Index
- Article 90—Introduction
- Article 100—Definitions
- Article 110—Requirements for Electrical Installations
- Article 300—Wiring Methods

We will refer to the rest of the *Code* as needed; but these are general articles, and we use them continually as applicable.

You must study these parts of the *Code* thoroughly several times until you have a complete understanding of the content in these articles if you expect to successfully pass your examination.

Introduction

Let us turn to Article 90. We often pass over the Introduction of many books that we read or study. The introduction is an important part of any book. It is the application guidelines. Did you read Article 90? Read Article 90 several times. It is short, only about four pages, but it is a very important part of the document, especially for someone about to embark on the study of the *National Electrical Code*®.

Article 90 covers the foundation requirements of the *NEC*®. The purpose is the safeguarding of people and property from the hazards that can occur while using electricity. The *Code* contains rules that, when complied with, along with proper maintenance, will result in an installation that is essentially free from hazards. But it may not be efficient, convenient, or adequate for good service. The requirements are minimal and generally do not provide capacity for future expansion of the electrical system.

FPN: Hazards often occur because of overloading of wiring systems by methods or usage not in conformity with this Code. *This occurs because initial wiring did not provide for increases in the use of electricity. An initial adequate installation and reasonable provisions for system changes provide for future increases in the use of electricity.*

Reprinted with permission from NFPA 70-2005.

The *NEC*® is not a design specification, but it is a design manual. For example, specific manufacturers' products are not specified; however, no responsible designer would begin to design an electrical system without considering minimal *Code* requirements such as load calculations, service size and location, and grounding.

The *NEC*® is not an instruction manual for untrained persons.

Contained in the Introduction is the scope, and it is imperative that the user know the bounds of the *NEC*® (90.2(A)).

90.2 Scope. (A) Covered. *This* Code *covers the installation of electrical conductors, equipment, and raceways; signaling and communications conductors, cables, equipment, and raceways; and optical fiber cables and raceways for the following:*

(1) Public and private premises, including buildings, structures, mobile homes, recreational vehicles, and floating buildings

(2) Yards, lots, parking lots, carnivals, and industrial substations

    FPN to (2): For additional information concerning such installations in an industrial or multibuilding complex, see ANSI C2-2002, National Electrical Safety Code.

(3) Installations of conductors and equipment that connect to the supply of electricity

---

(4) Installations used by the electric utility, such as office buildings, warehouses, garages, machine shops, and recreational buildings that are not an integral part of a generating plant, substation, or control center.

Reprinted with permission from NFPA 70-2005.

This Code does not cover (90.2(B)):

(B) Not Covered. *This* Code *does not cover the following:*

(1) Installations in ships, watercraft other than floating buildings, railway rolling stock, aircraft, or automotive vehicles other than mobile homes and recreational vehicles

    FPN: Although the scope of this Code *indicates that the* Code *does not cover installations in ships, portions of this* Code *are incorporated by reference into Title 46,* Code of Federal Regulations, *Parts 110–113.*

(2) Installations underground in mines and self-propelled mobile surface mining machinery and its attendant electrical trailing cable

(3) Installations of railways for generation, transformation, transmission, or distribution of power used exclusively for operation of rolling stock or installations used exclusively for signaling and communications purposes

(4) Installations of communications equipment under the exclusive control of communications utilities located outdoors or in building spaces used exclusively for such installations

(5) Installations under the exclusive control of an electric utility where such installations

    a. Consist of service drops or service laterals and associated metering, or

    b. Are located in legally established easements, rights-of-way, or by other agreements either designated by or recognized by public service commissions, utility commissions, or other regulatory agencies having jurisdiction for such installations, or

    c. Are on property owned or leased by the electric utility for the purpose of communications, metering, generation, control, transformation, transmission, or distribution of electric energy.

    FPN to (4) and (5): Examples of utilities may include those entities that are typically designated or recognized by governmental law or regulation by public service/utility commissions and that install, operate, and maintain electric supply (such as generation, transmission, or distribution systems) or communication systems (such as telephone, CATV, Internet, satellite, or data services). Utilities may be subject to compliance with codes and standards covering their regulated activities as adopted under governmental law or regulation. Additional information can be found through consultation with the appropriate governmental bodies, such as state regulatory commissions, Federal Energy Regulatory Commission, and Federal Communications Commission.

Reprinted with permission from NFPA 70-2005.

Included in 90.3 is the arrangement of the *Code* and hierarchy. The *Code* is divided into an Introduction, nine chapters, and annexes A, B, and C.

- Chapters 1, 2, 3, and 4 apply generally and are applicable to all installations—branch circuitry, motors, appliances, lighting.

- Chapters 5, 6, and 7 apply to special occupancies—places of assembly, hazardous locations, agricultural buildings, mobile homes, health care facilities; special equipment—swimming pools, signs, elevators, welders; or other special conditions—emergency systems, fire alarms, power-limited circuits. Although the first four still apply, these chapters augment or modify the general rules for the particular conditions.

Example: Article 514 covers the hazardous requirements for wiring a service station. It does not cover other areas within that service station facility. It does not cover other electrical needs, such as service, the grounding, the branch-circuits, or the feeders, and more importantly it does not cover other environmental concerns, such as corrosion or wet locations. The designer must take all these concerns and any others that apply and design the installation so that it will provide good service, have adequate capacity, and provide all the safety needed for the normal operation throughout the life of the facility.

- Chapter 8 is a stand-alone chapter and covers communications systems; it is independent of the other chapters except where they are specifically referenced.

The *NEC®* is divided into the introduction and nine chapters, as shown here. Chapters 1, 2, 3, and 4 apply generally; Chapters 5, 6, and 7 apply to special occupancies, special equipment, or other special conditions. These latter chapters supplement or modify the general rules. Chapters 1 through 4 apply except as amended by Chapters 5, 6, and 7 for the particular conditions.

Chapter 8 covers communications systems and is not subject to the requirements of Chapters 1 through 7 except where the requirements are specifically referenced in Chapter 8.

Chapter 9 consists of tables.

Annexes are not part of the requirements of the Code, but are included for informational purposes only.

Reprinted with permission from NFPA 70-2005.

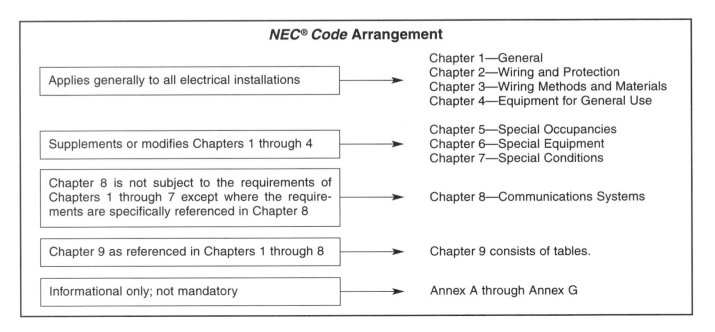

NEC® Code Arrangement

Applies generally to all electrical installations →
Chapter 1—General
Chapter 2—Wiring and Protection
Chapter 3—Wiring Methods and Materials
Chapter 4—Equipment for General Use

Supplements or modifies Chapters 1 through 4 →
Chapter 5—Special Occupancies
Chapter 6—Special Equipment
Chapter 7—Special Conditions

Chapter 8 is not subject to the requirements of Chapters 1 through 7 except where the requirements are specifically referenced in Chapter 8 →
Chapter 8—Communications Systems

Chapter 9 as referenced in Chapters 1 through 8 →
Chapter 9 consists of tables.

Informational only; not mandatory →
Annex A through Annex G

- Chapter 9 consists of tables and examples.
- Annex A lists extract documents; see 90.3 for an explanation.
- Annex B is for information only and contains formula application information.
- Annex C contains optional wire fill tables.

The Introduction lets us know that the *Code* is intended as suitable for mandatory application by those adopting it—that is, governmental bodies exercising legal jurisdiction over electrical installations and for use by insurance inspectors. It also places the responsibility for enforcement on the authority having jurisdiction (AHJ).

NEC® 90.5 gives the style of the document and covers the mandatory rules, which are characterized by the word "**shall**" and explains that explanatory material appears in the form of fine print notes, which are characterized by the designation "**FPN**."

90.6 Formal Interpretations. *To promote uniformity of interpretation and application of the provisions of this* Code, *formal interpretation procedures have been established and are found in the NFPA Regulations Governing Committee Projects.*
Reprinted with permission from NFPA 70-2005.

Warning: The commentary found in the NFPA *National Electrical Code Handbook* is **not** a formal interpretation of the *NEC®*.

NEC® 90.7 states that factory-installed internal wiring or the construction of equipment need not be inspected at the time of installation of the equipment, except to detect alterations or damage, if the equipment has been listed by a qualified electrical testing laboratory that is recognized and that requires suitability for installation in accordance with this *Code*.

See Examination, Identification, Installation, and Use of Equipment, 110.3.
See definition of "Listed," Article 100.

90.9(A) Units of Measurement. *For the purpose of this* Code, *metric units of measurement are in accordance with the modernized metric system known as the International System of Units (SI).*
Reprinted with permission from NFPA 70-2005.

Values of measurement in the *Code* text will be followed by an approximate equivalent value in SI

---

units. Tables will have a footnote for SI conversion units used in the table.

Conduit size, wire size, horsepower designation for motors, and trade sizes that do not reflect actual measurements—for example, box sizes—will not be assigned dual designation SI units.

NEC® Units of Measurement (90.9)

- In the *NEC®*, metric units of measurement are in accordance with the modernized metric system known as the International System of Units (SI).
- The SI units appear first, followed by the inch-pound units in parentheses. The conversion from the inch-pound units to SI units is based on hard conversion, except as provided here.

(C)(1) **Trade Sizes.** *Where the actual measured size of a product is not the same as the nominal size, trade size designators shall be used rather than dimensions. Trade practices shall be followed in all cases.*

Example:

Table 300.1(C) Metric Designator and Trade Sizes

Metric Designator	Trade Size
12	3/8
16	1/2
21	3/4
27	1
35	1 1/4
41	1 1/2
53	2
63	2 1/2
78	3
91	3 1/2
103	4
129	5
155	6

Note: The metric designators and trade sizes are for identification purposes only and are not actual dimensions.

(C)(2) **Extracted Material.** *Where material is extracted from another standard, the context of the original material shall not be compromised or violated. Any editing of the extracted text shall be confined to making the style consistent with that of the* NEC.

(C)(3) **Industry Practice.** *Where industry practice is to express units in inch-pound units, the inclusion of SI units shall not be required.*

Example: **Extracted Text to Article 516.3 from NFPA 33 and NFPA 34**

(4) For open dipping and coating operations, all space within a 1.5-m (5-ft) radial distance from the vapor sources extending from these surfaces to the floor. The vapor source shall be the liquid exposed in the process and the drainboard, and any dipped or coated object from which it is possible to measure vapor concentrations exceeding 25 percent of the lower flammable limit at a distance of 305 mm (1 ft), in any direction, from the object.

(5) Sumps, pits, or belowgrade channels within 7.5 m (25 ft) horizontally of a vapor source. If the sump, pit, or channel extends beyond 7.5 m (25 ft) from the vapor source, it shall be provided with a vapor stop or it shall be classified as Class I, Division 1 for its entire length. *

(C)(4) **Safety.** *Where a negative impact on safety would result, soft conversion shall be used.*

Example: **NEC® 110.31(A) Live Parts (over 600 volts) Guarded Against Accidental Contact.**
The distance from the fence to live parts shall be not less than the following:

601 to 13,799 volts	10 ft (3.05 m)
13,800 to 230,000 volts	15 ft (4.57 m)
Over 230,000 volts	18 ft (5.49 m)

(D) **Compliance.** *The conversion from inch-pound units to SI units shall be permitted to be an approximate conversion. Compliance with the numbers shown in either the SI system or the inch-pound system shall constitute compliance with this* Code. *

NEC® 90.9 Units of Measurement
FPN No. 1: Hard conversion is considered a change in dimensions or properties of an item into new sizes that might or might not be interchangeable with the sizes used in the original measurement. Soft conversion is considered a direct mathematical conversion and involves a change in the description of an existing measurement but not in the actual dimension.
FPN No. 2: SI conversions are based on IEEE/ASTM SI 10-1997, Standard for the Use of the International System of Units (SI): The Modern Metric System. *

* Reprinted with permission from NFPA 70-2005.

As you can see, Article 90 is very important, and one could never apply the *Code* correctly without first having a thorough understanding of the Introduction, Article 90.

General Requirements

You must continually refer to the definitions, *NEC®* Article 100, as you apply the requirements of the *NEC®*. These terms are unique and essential to this document, which is the reason they have been included in this document.

Article 110 covers the general requirements. *NEC®* 110.3 tells us that we must select equipment and material suitable for the installation, environment, and/or application, and that it must be installed in accordance with any instructions included with its listing or labeling. *NEC®* 110.9 and 110.10 give us some very strong mandatory requirements on interrupting rating and circuit impedance that we cannot ignore. If the equipment is intended to break current, it must be rated to interrupt the available current at fault conditions, and the system voltage and the circuit components including the impedance shall be coordinated so that the circuit protective device will clear any fault without excessive damage to the equipment. *NEC®* 110.8 and 110.12 state that only recognized methods are permitted and that we are required to make our installation in a neat and professional manner. *NEC®* 110.14 gives us mandatory requirements for making splices and terminations. *NEC®* 110.14(C) states that the terminations must be considered when determining the ampacity of a circuit. Read this *NEC®* section very carefully; similar requirements have appeared in the UL Green and White books.

NEC® 110.26 states that we are required to carefully select the proper location for our 600 volts or less equipment (for equipment rated at over 600 volts, the *Code* references are *NEC®* Article 110 Parts III and IV, 110.30 through 110.59) so that proper work space can be maintained. This often requires us to check the building and mechanical plans so that encroachment can be avoided. To discover this encroachment in the last stages of the construction can be and often is very costly to one or more of the craft contractors. Other sections in Chapter 1 are equally important and must be continually referenced every time the *NEC®* is applied.

The title to Article 110 Part 5 has changed; however, the requirements have not changed. The new expanded title—Manholes and Other Electric Enclosures Intended for Personnel Entry, All Voltages—

clarifies that all enclosures that are intended for personnel entry shall be of sufficient size to maintain safe work space and clearances from live parts.

Wiring and Protection

Now that we have thoroughly studied the general requirements, the next step is to take a look at the wiring and protection requirements. Turn to Chapter 2 of the *National Electrical Code®*, Article 210—Branch Circuits.

Note: Many of the requirements in this article apply only to dwellings; read the pertinent sections very carefully before applying the requirements.

Note: Although, generally, extra circuits are not necessary, careful arrangement of the required branch-circuits is necessary for compliance with 210.11(A), (B), and (C).

You will also find specific branch-circuit requirements in other articles of the *NEC®*, such as Article 430 for Motors, Article 600 for Signs, and Article 517 for Health Care Facilities. *NEC®* 210.52 gives the guidelines necessary to avoid the use of extension cords in dwellings. 210.52(A) requires outlets to be located so that no point along the floor line is more than 6 feet from an outlet and wall spaces 2 feet wide have an outlet in them. This requirement ensures that it is not necessary to lay cords across door openings to reach an outlet. This section also points out that outlets that are a part of luminaires (light fixtures) and appliances, located within cabinets, or above 5½ feet (1.7 m) cannot be counted as one of those required. *NEC®* 210.52(A) through (H) gives the requirements for outlets including the appliance outlets and the ground-fault circuit protection where required. (Several of the subsections in 210.52 reference 210.8, which gives the requirements for GFCIs in dwellings and other locations.) Now look at *NEC®* 210.70(A) for the required lumination (lighting) outlets and wall switches. Extra lumination (lighting) can be added for the customer, but the required lighting will provide a safe illumination. *NEC®* 210.70(A) generally requires a switch-controlled light in each room. Stairways are required to be lighted with a switch at the top and bottom of the stairs. *NEC®* 210.70(A) now requires a switch-controlled lighting outlet in attics and under floor spaces where used for storage or equipment that requires servicing. This section also requires a switch-controlled light at outside exits or entrances. After we

have established the branch-circuit requirements, we can do the general lighting load calculations in accordance with Table 220.12 and footnotes. (See Figure 4–1.) The total calculated load cannot be ascertained until all of the equipment loads are known and added to the general-purpose branch-circuits. Chapter 2, Articles 215 and 225, will also guide us as we select and size our inside and outside feeders.

Article 225 for outside branch-circuits and feeders has been revised in the past two *NEC®* editions, and Part II provides a mirror image of the service requirements for "more than one building or structure on the same property under single management." The requirements for the feeder supplying that building or structure from the service are almost identical to those for a service. *NEC®* 225.30 limits the building to be supplied by one feeder or branch-circuit except for special conditions such as fire pumps and emergency circuits and for special occupancies. These are requirements similar to those of *NEC®* 230.2. *NEC®* 225.31 through 225.39 cover the disconnect requirements. *NEC®* 225.40 requires ready access to the overcurrent devices. Part III covers the requirements for over-600-volt feeders.

Article 230—Services is the logical next step and also the next article. We are beginning to see some rationale for the chapter and article arrangement and layout. The diagram 230.1 (page 51) is an excellent aid for using this article as the sections apply to the installation. Although materials are not specified, the methods and requirements are specific.

Note: When designing the service, or solving questions related to the service, it is important that you properly define the components of the installation (see Article 100—Definitions). (See Figure 4–2 [page 52].)

NEC® 230.42(A) requires the service-entrance conductors to be of sufficient size to carry the loads as calculated in Article 220.

Article 240 covers the overcurrent protection for conductors requirements. This part of the *Code* has some very specific rules that will apply to all areas in the *Code*. For example, Parts I through VII of Article 240 cover the general requirements for overcurrent protection and overcurrent protective devices 600 volts or less. Part VIII covers overcurrent protection for those portions of supervised industrial installations operating at voltages of not more than 600 volts, nominal. Part IX covers overcurrent protection over 600 volts, nominal.

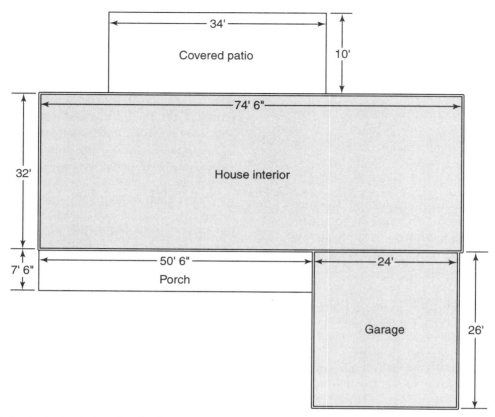

Figure 4–1 A typical single-family dwelling with an attached garage and covered porch on the front with a covered patio in the back. Neither the porch nor the patio is enclosed.

Example:

In Figure 4-1, what is the general lighting load for this house?

    A. 11,181 VA

    B. 11,180.25 VA

    C. 7152 VA

    D. 9024 VA

Answer: C. *NEC®* 220.12 states that the computed area does not include porches and garages in *NEC®* Table 220.12. The general lighting load for a dwelling unit[a] is 3 volt-amperes per square foot.

The superscript [a] directs us to a note at the bottom of Table 220.12 that references *NEC®* 220.14(J), which states that no additional load calculations are required.

> FPN: Overcurrent protection for conductors and equipment is provided to open the circuit if the current reaches a value that will cause an excessive or dangerous temperature in conductors or conductor insulation. See also 110.9 for requirements for interrupting ratings and 110.10 for requirements for protection against fault currents.

NEC® 240.2 was new to the 2002 *NEC®*. This section contains special definitions that apply to this article of the *Code*.

230.6 Conductors Considered Outside the Building. Conductors shall be considered outside of a building or other structure under any of the following conditions:

(1) Where installed under not less than 50 mm (2 in.) of concrete beneath a building or other structure

(2) Where installed within a building or other structure in a raceway that is encased in concrete or brick not less than 50 mm (2 in.) thick

(3) Where installed in any vault that meets the construction requirements of Article 450, Part III

(4) Where installed in conduit and under not less than 450 mm (18 in.) of earth beneath a building or other structure.

Coordination (Selective). Localization of an overcurrent condition to restrict outages to the circuit or equipment affected, accomplished by the choice of overcurrent protective devices and their ratings or settings.

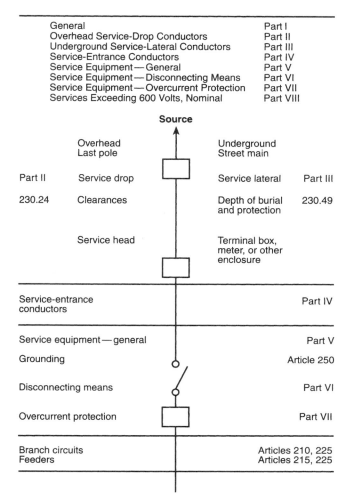

General		Part I
Overhead Service-Drop Conductors		Part II
Underground Service-Lateral Conductors		Part III
Service-Entrance Conductors		Part IV
Service Equipment—General		Part V
Service Equipment—Disconnecting Means		Part VI
Service Equipment—Overcurrent Protection		Part VII
Services Exceeding 600 Volts, Nominal		Part VIII

Source

	Overhead Last pole		Underground Street main	
Part II	Service drop		Service lateral	Part III
230.24	Clearances		Depth of burial and protection	230.49
	Service head		Terminal box, meter, or other enclosure	

Service-entrance conductors	Part IV
Service equipment—general	Part V
Grounding	Article 250
Disconnecting means	Part VI
Overcurrent protection	Part VII
Branch circuits	Articles 210, 225
Feeders	Articles 215, 225

Figure 230.1 Services.

(Reprinted with permission from NFPA 70-2005, the *National Electrical Code®* Copyright © 2004–, National Fire Protection Association, Quincy, MA 02269. This reprinted material is not the complete and official position of the National Fire Protection Association on the referenced subject, which is represented only by the standard in its entirety.)

Current-Limiting Overcurrent Protective Device. *A device that, when interrupting currents in its current-limiting range, reduces the current flowing in the faulted circuit to a magnitude substantially less than that obtainable in the same circuit if the device were replaced with a solid conductor having comparable impedance.*

Supervised Industrial Installation. *For the purposes of Part VIII, the industrial portions of a facility where all of the following conditions are met:*

---

(1) Conditions of maintenance and engineering supervision ensure that only qualified persons will monitor and service the system.

(2) The premises wiring system has 2500 kVA or greater of load used in industrial process(es), manufacturing activities, or both, as calculated in accordance with Article 220.

(3) The premises has at least one service that is more than 150 volts to ground and more than 300 volts phase-to-phase.

This definition excludes installations in buildings used by the industrial facility for offices, warehouses, garages, machine shops, and recreational facilities that are not an integral part of the industrial plant, substation, or control center.

Tap Conductors. *As used in this article, a tap conductor is defined as a conductor, other than a service conductor, that has overcurrent protection ahead of its point of supply that exceeds the value permitted for similar conductors that are protected as described elsewhere in 240.4.*

* Reprinted with permission from NFPA 70-2005.

240.4(E) Tap Conductors. *Tap conductors shall be permitted to be protected against overcurrent in accordance with the following:*

(1) 210.19(A)(3) and (A)(4) Household Ranges and Cooking Appliances and Other Loads
(2) 240.5(B)(2) Fixture Wire
(3) 240.21 Location in Circuit
(4) 368.17(B) Reduction in Ampacity Size of Busway
(5) 368.17(C) Feeder or Branch Circuits (busway taps)
(6) 430.53(D) Single Motor Taps *

240.21 Location in Circuit. *Overcurrent protection shall be provided in each ungrounded circuit conductor and shall be located at the point where the conductors receive their supply except as specified in 240.21(A) through (G). No conductor supplied under the provisions of 240.21(A) through (G) shall supply another conductor under those provisions, except through an overcurrent protective device meeting the requirements of 240.4.* *

Note: This text in *NEC®* 240.21 clarifies that you can never tap a tap! (See Figure 4–3.)

NEC® 240.3 contains a list of about 40 *NEC®* articles that contains specific overcurrent requirements

* Reprinted with permission from NFPA 70-2005.

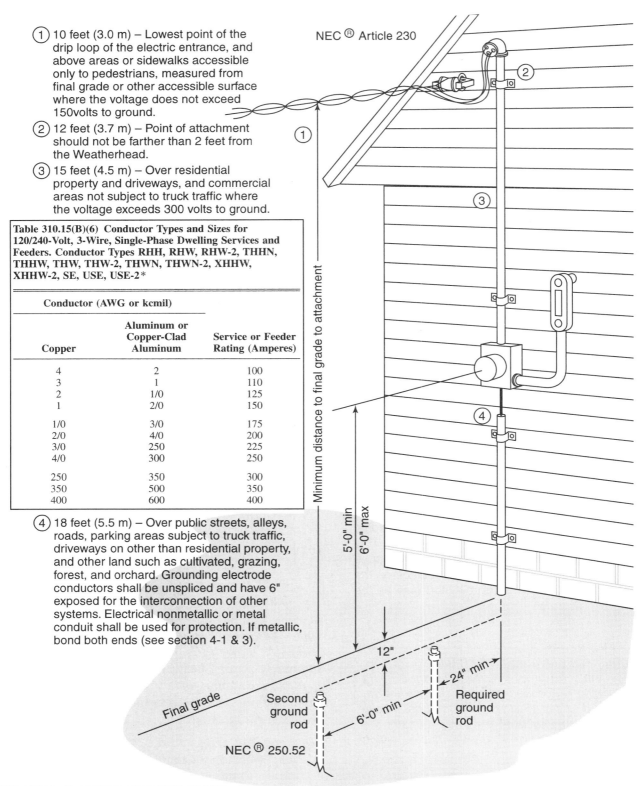

① 10 feet (3.0 m) – Lowest point of the drip loop of the electric entrance, and above areas or sidewalks accessible only to pedestrians, measured from final grade or other accessible surface where the voltage does not exceed 150volts to ground.

② 12 feet (3.7 m) – Point of attachment should not be farther than 2 feet from the Weatherhead.

③ 15 feet (4.5 m) – Over residential property and driveways, and commercial areas not subject to truck traffic where the voltage exceeds 300 volts to ground.

NEC ® Article 230

Table 310.15(B)(6) Conductor Types and Sizes for 120/240-Volt, 3-Wire, Single-Phase Dwelling Services and Feeders. Conductor Types RHH, RHW, RHW-2, THHN, THHW, THW, THW-2, THWN, THWN-2, XHHW, XHHW-2, SE, USE, USE-2*

Conductor (AWG or kcmil)		
Copper	Aluminum or Copper-Clad Aluminum	Service or Feeder Rating (Amperes)
4	2	100
3	1	110
2	1/0	125
1	2/0	150
1/0	3/0	175
2/0	4/0	200
3/0	250	225
4/0	300	250
250	350	300
350	500	350
400	600	400

④ 18 feet (5.5 m) – Over public streets, alleys, roads, parking areas subject to truck traffic, driveways on other than residential property, and other land such as cultivated, grazing, forest, and orchard. Grounding electrode conductors shall be unspliced and have 6" exposed for the interconnection of other systems. Electrical nonmetallic or metal conduit shall be used for protection. If metallic, bond both ends (see section 4-1 & 3).

* Reprinted with permission from NFPA 70-2005.

Figure 4-2 An example of a simple electrical service supplying a building. (1) Vertical clearing requirements; (2) Point of attachment should be located to remove stress on the mast; (3) *NEC*® Table 310.15(B)(6) [reprinted with permission from NFPA 70-2005]); (4) Example of grounding electrodes as required by *NEC*® 250.50 and 250.52.

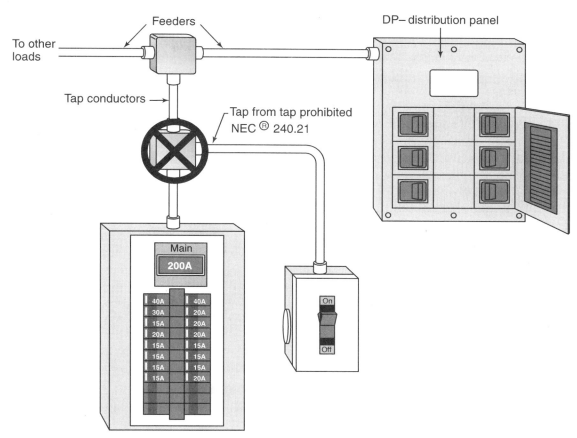

Figure 4–3 This drawing depicts a violation of the 2005 *NEC®*. The *Code* permits smaller conductor taps from a feeder to supply more than one piece of equipment. However, the tap from the tap conductors feeding the panelboard to supply the switch is not allowed. You are not permitted to "tap a tap." (See 240.2 Definition of a Tap, 240.4 Protection of Conductors, and 240.21.)

for systems and equipment such as emergency systems, motors generators, and the like, that are in addition to Article 240.

NEC® 240.4 covers the protection of conductors and states that other than flexible cords and fixture wires, all conductors shall be protected against overcurrent in accordance with their ampacities specified in *NEC®*310.15, unless otherwise permitted or required in 240.4(A) through 240.4(G).

Example: Conductor overload protection is not required for which of the following where the interruption of the circuit would create a hazard?
- A. Material-handling magnet
- B. Motors
- C. Lighting
- D. All of these

Answer: A. *NEC®* covers power loss hazards such as a material-handling magnet. If there was a power

---

loss while the magnet was moving a heavy object, the object would fall and possibly cause a severe injury.

Example: For overcurrent devices 800 amperes or less, the next higher overcurrent device is always permitted. True or False?

Answer: False. Although this is permitted where all of the conditions in *NEC®* 240.4(B) are met. It is not always permitted.

Example: For overcurrent devices 800 amperes or more, the next higher overcurrent device is never permitted. True or False?

Answer: True. *NEC®* 240.4(C) requires the overcurrent rating to be greater than the ampacity of the conductors it protects.

Small Size Conductor Overcurrent Protection Rules

The overcurrent protection ratings (circuit breaker or fuse) cannot exceed 15 amperes for 14 AWG copper conductor, 20 amperes for 12 AWG copper conductor, and 30 amperes for 10 AWG copper conductor; or 15 amperes for 12 AWG and 25 amperes for 10 AWG aluminum and copper-clad aluminum conductor after any correction factors for ambient temperature and number of conductors have been applied except as permitted in *NEC®* 240.4(E) through 240.4(G).

Example: A 120-volt, single-phase circulating hot water pump in a multifamily dwelling is supplied by a 12-2 W/Grd Type NM cable. What is the maximum size overcurrent circuit breaker or fuse protection for this circuit?

    A. 10 ampere
    B. 20 ampere
    C. 25 ampere
    D. None of these

Answer: D. There is not enough information to answer this question. *NEC®* 240.3 states that motor overcurrent protection is governed by Article 430 Part III.

NEC® 240.4(E) permits tap conductors to be protected in accordance with the applicable *Code* section. For example, the rules for 25-foot tap conductors per 240.21(B)(2) would require the tap conductors to be protected from physical damage and have an ampacity of not less than one third the rating of the overcurrent device protecting the feeder conductors terminating in a single circuit breaker or a single set of fuses that limit the load to the ampacity of the tap conductors. The device can supply any number of additional overcurrent devices on its load side.

Example: A 3 AWG THWN feeder protected by a 100-ampere circuit breaker is tapped in a junction box with _____ AWG THWN conductors 20 feet (6 m) long to feed a 30-ampere single set of fuses. What is the minimum size THWN conductor permitted to be used for making this tap?

    A. 3 AWG
    B. 6 AWG
    C. 8 AWG
    D. 10 AWG

Answer: D. *NEC®* 240.21(B)(2) would require 1/3 of 100 or 33.34-ampere wire.

Table 310.16
- 3 AWG has an allowable ampacity of 100 amperes.
- 10 AWG has an allowable ampacity of 35 amperes.

See 2005 *NEC®* 240.4 Protection of Conductors:
(A) Power Loss Hazard
(B) Devices Rated 800 Amperes or Less
(C) Devices Rated Over 800 Amperes
(D) Small Conductors
(E) Tap Conductors
(F) Transformer Conductors
(G) Overcurrent Protection for Specific Conductor Applications

NEC® 240.21 covers the tap rules as they are permitted throughout the *Code*. Experts generally agree that in all cases the conductors extending from the secondary of a transformer to the overcurrent device is a tap. They also agree that you can never tap a tap.

See 2005 *NEC®* 240.21 Location in Circuit:
(A) Branch-Circuit Taps
(B) Feeder Taps
(C) Transformer Secondary Conductors
(D) Service Conductors
(E) Busway Taps
(F) Motor Circuit Taps
(G) Conductors from Generator Terminals

This article also contains guidance for fuses and circuit breakers, their application, and some manufacturing requirements. For instance, 240.60(B) regulates current-limiting fuses. They must be manufactured so that they are different from standard fuses: equipment designed or modified to accept current-limiting fuses will reject standard fuses. Manufacturers accomplish this by placing a notch in the load side (bottom) of the fuse blade, and equipment designed to accept them has a pin in the load side fuse holder. So current-limiting types can be installed, but standard types cannot. *NEC®* 240.83 gives similar specifics for circuit breakers.

Article 240 is a general use article and must be referred to on all installations. Part VIII covers those portions of supervised industrial installations operating at voltages not exceeding 600 volts nominal. *NEC®* 240 Part VIII applies only to large industrial facilities and only to those portions of the electrical system in the supervised industrial installation used exclusively for manufacturing or process control activities.

NEC® 240.2 defines supervised installations as the industrial portions of a facility where all conditions are met. This definition makes it clear that it does not apply to buildings used as offices, warehouses, garages, and so on, that are not part of the industrial plant, substation, or control center.

Part IX covers overcurrent protection over 600 volts nominal. This information was previously located in the 1996 *NEC®* Article 710. Article 710 was deleted, and all requirements for above-600-volt installations were relocated into the appropriate article.

NEC® Article 250—Grounding

The last major article in Chapter 2 is Article 250. (I did not forget Article 280—Surge Arresters and Article 285—Transient Voltage Surge Suppressors: TVSS. If you have a question related to surge arresters or TVSS, do not forget these articles.) Grounding is one of the very important parts of any electrical installation project and will be an important part of your studies. Article 250 is very specific. As you design the grounding system for an installation, you will find it to be concise and complete for most installations. However, specific grounding requirements are referenced in several other articles of the *NEC®*. If you are researching a question related to grounding, you may find the answer in the article specifically addressing the question. This is an article that you will become very familiar with during your studies of the *Code*; however, you may find it more efficient to use the index to see all references and where they are located in the *Code*.

NEC® 250.4(A)(5) makes it clear that in all cases the fault-current path must be electrically continuous, permanent, and capable of carrying the maximum fault likely to be imposed on it. It must have sufficiently low impedance to facilitate the operation of the overcurrent device(s) under fault conditions. This section goes hand in hand with *NEC®* 110.10, which covers all components of a system.

NEC® 110.10 Circuit Impedance and Other Characteristics. *The overcurrent protective devices, the total impedance, the component short-circuit current ratings, and other characteristics of the circuit to be protected shall be selected and coordinated to permit the circuit-protective devices used to clear a fault to do so without extensive damage to the electrical components of the circuit. This fault shall be assumed to be either between two or more of the circuit conductors, or between any circuit con-*

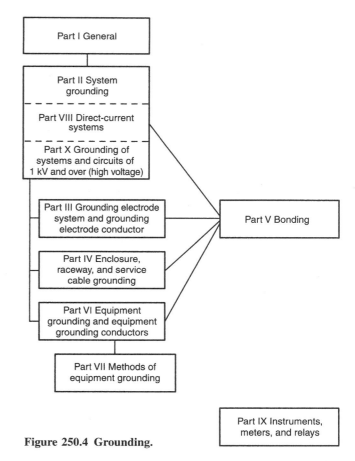

Figure 250.4 Grounding.

(Reprinted with permission from NFPA 70-2005, the *National Electrical Code®* Copyright © 2004–, National Fire Protection Association, Quincy, MA 02269. This reprinted material is not the complete and official position of the National Fire Protection Association on the referenced subject, which is represented only by the standard in its entirety.)

---

ductor and the grounding conductor or enclosing metal raceway. Listed products applied in accordance with their listing shall be considered to meet the requirements of this section.

Reprinted with permission from NFPA 70-2005.

NEC® 250.24 clearly lays out the requirements for system grounding.

(1) General. *The connection shall be made at any accessible point from the load end of the service drop or service lateral to and including the terminal or bus to which the grounded service conductor is connected at the service disconnecting means.*

(C)(1) Routing and Sizing. *This conductor shall be routed with the phase conductors and shall not be smaller than the required grounding electrode conductor* *specified in Table 250.66 but shall not be required to be larger than the largest ungrounded service-entrance phase conductor. In addition, for service-entrance phase conductors larger than 1100 kcmil copper or 1750 kcmil aluminum, the grounded conductor shall not be smaller than 12½ percent of the area of the largest service-entrance phase conductor. The grounded conductor of a 3-phase, 3-wire delta service shall have an ampacity not less than that of the ungrounded conductors.*

A premises wiring system supplied by a grounded ac service shall have a grounding electrode conductor connected to the grounded service conductor at each service in accordance with the provisions of 250.24 (A)(1) through (A)(5).

***NEC*® 250.24(D) Grounding Electrode Conductor.** *A grounding electrode conductor shall be used to connect the equipment grounding conductors, the service-equipment enclosures, and, where the system is grounded, the grounded service conductor to the grounding electrode(s) required by Part III of this article.*

High-impedance grounded neutral system connections shall be made as covered in 250.36.

Reprinted with permission from NFPA 70-2005.

NEC® 250.32 makes it clear that the neutral shall not be regrounded at the second building or structure if there are any common metallic paths, such as water line, gas line, telephone, CATV, or process piping. This means that, in most cases, an equipment grounding conductor (metal raceway) must be run with the circuit conductors.

NEC® Article 250 Part III has been rewritten with significant changes for the accepted and required grounding electrode system.

NEC® 250.50 now requires that in each building or structure all the electrodes as described in 250.52 (A)(1) through 250.52(A)(6) present must be bonded together to form the grounding electrode system. The significance of this is that the concrete-encased (rebar) electrode is now required to be a part of the electrode system unless the footings are poured and the rebar is not accessible.

Where none of the electrodes covered in 250.52(A)(1) through 250.52(6) exists, then one of the electrodes described in 250.52(A)(4) through 250.52(A)(7) must be installed and used.

---

* Reprinted with permission from NFPA 70-2005.

- *NEC*® 250.52(A) lists the types of electrodes permitted such as underground water pipe in direct contact with earth 10 feet (3.0 m) or more, effectively grounded building steel, concrete-encased electrodes, and the like.

- *NEC*® 250.52(B) lists the two types of electrodes not permitted:

  (1) Metal underground gas piping system
  (2) Aluminum electrodes

- *NEC*® 250.53(A) through (H) covers the installation requirements for grounding electrodes.

  Note: Unless you test the resistance of the ground rod and it has 25 ohms or less, you must drive two electrodes to supplement the water pipe.

- *NEC*® 250.54 covers supplementary grounding electrodes.

- *NEC*® 250.56 covers resistance requirements for rod, pipe, and plate grounding electrodes.

- *NEC*® 250.58 covers "common" grounding electrodes.

- *NEC*® 250.60 covers the rules for the use of "air terminals."

- *NEC*® 250.62 covers the grounding electrode conductor material. *The material selected shall be resistant to any corrosive condition existing at the installation.*

- *NEC*® 250.64(A) through (F) covers the installation requirements for grounding electrode conductors.

- *NEC*® 250.66(A) through (C) and Table 250.66 cover the size of alternating current (ac) grounding electrode conductors.

- *NEC*® 250.68(A) and (B) covers the grounding electrode conductor and bonding jumper connection to the grounding electrode requirements.

- *NEC*® 250.70 covers the rules for making the connection of the grounding conductor to the electrode.

Now that we have our branch circuitry, the feeders, the service, the overcurrent protection, and the grounding designed into the installation, it is time to select the wiring methods and materials; those can all be found in Chapter 3 of the *NEC*®.

Wiring Methods

Chapter 3 was editorially revised in the 2002 *NEC*®. All articles were renumbered and relocated.

Chapter 3 Article Numbers and Acronyms

Article 300 Wiring Methods
Article 310 Conductors For General Wiring
Article 312 Cabinets, Cutout Boxes and Meter Socket Enclosures
Article 314 Outlet, Device, Pull and Junction Boxes, Conduit Bodies and Fittings
Article 320 Armored Cable **Type AC**
Article 322 Flat Cable Assemblies **Type FC**
Article 324 Flat Conductor Cable **Type FCC**
Article 326 Integrated Gas Spacer Cable **Type IGS**
Article 328 Medium Voltage Cable **Type MV**
Article 330 Metal-Clad Cable **Type MC**
Article 332 Mineral-Insulated Metal-Sheathed Cable **Type MI**
Article 334 Nonmetallic-Sheathed Cable **Types NM, NMC, and NMS**
Article 336 Power and Control Tray Cable **Type TC**
Article 338 Service Entrance Cable **Types SE and USE**
Article 340 Underground Feeder and Branch-Circuit Cable **Type UF**
Article 342 Intermediate Metal Conduit **Type IMC**
Article 344 Rigid Metal Conduit **Type RMC**
Article 348 Flexible Metal Conduit **Type FMC**
Article 350 Liquidtight Flexible Metal Conduit **Type LFMC**
Article 352 Rigid Nonmetallic Conduit **Type RMC**
Article 353 High Density Polyethylene Conduit **Type HDPE**
Article 354 Nonmetallic Underground Conduit with Conductors **Type NUCC**
Article 356 Liquidtight Flexible Nonmetal Conduit **Type LFNC**
Article 358 Electrical Metallic Tubing **Type EMT**
Article 360 Flexible Metallic Tubing **Type FMT**
Article 362 Electrical Nonmetallic Tubing **Type ENT**
Article 366 Auxiliary Gutters
Article 368 Busways
Article 370 Cablebus
Article 372 Cellular Concrete Floor Raceways
Article 374 Cellular Metal Floor Raceways
Article 376 Metal Wireways
Article 378 Nonmetallic Wireways
Article 380 Multioutlet Assembly
Article 382 Nonmetallic Extensions

[continued]

Article 384 Strut-Type Channel Raceway
Article 386 Surface Metal Raceways
Article 388 Surface Nonmetallic Raceways
Article 390 Underfloor Raceways
Article 392 Cable Trays
Article 394 Concealed Knob and Tube Wiring
Article 396 Messenger Supported Wiring
Article 398 Open Wiring on Insulators

The individual article sections have been renumbered, using a common numbering system. For example, with only a few exceptions, all of the wiring methods in Chapter 3 will start with a "Scope" numbered as 3XX.1. This new common numbering system will make using the *Code* much quicker and easier. Once it is learned that "Securing and Supporting" requirements can be found in each Article 3XX.30 or that "Uses Not Permitted" can be found in 3XX.12, then the entire article will not need to be scanned to find these requirements.

Chapter 3 Wiring Methods Common Numbering

Part I General
3XX.1 Scope
3XX.2 Definitions
3XX.3 Other Articles
3XX.6 Listing Requirements

Part II Installation
3XX.10 Uses Permitted
3XX.12 Uses Not Permitted
3XX.14 Dissimilar Metals
3XX.16 Temperature Limits
3XX.20 Size
3XX.22 Number of Conductors
3XX.24 Bends—How made
3XX.26 Bends—Number in one run
3XX.28 Reaming, Threading and Trimming
3XX.30 Securing and Supporting
3XX.40 Boxes and Fittings
3XX.42 Couplings and Connectors
3XX.44 Expansion Fittings
3XX.46 Bushings
3XX.48 Joints
3XX.50 Conductor Terminations
3XX.56 Splices and Taps
3XX.60 Grounding

Part III Construction Specifications
3XX.100 Construction
3XX.110 Corrosion Protection
3XX.120 Marking
3XX.130 Standard Lengths

I hope you have studied Article 300. As I stated earlier, this article covers general requirements that must be applied to all the wiring methods. The wiring method selected must be appropriate for the environment, meet the physical protection requirements, temperature, and all other pertinent conditions that may affect the material selected.

> **Note:** The uses permitted and not permitted appear in each wiring method article. Restrictions of use vary. The answer to your question may be in those NEC® sections.

The condition or question may address the wiring method directly or may require you to select an approved method based on these factors. Be sure to study all of the "uses permitted" and the "uses not permitted" in each specific article carefully.

> For example, when selecting Rigid Metal Conduit (RMC), you only need to consider the corrosion elements (NEC® 300.6). However, when selecting Nonmetallic-Sheathed Cable (Type NM), many factors must be considered (300.4, 336.10, and 336.12). Once this has been done, a wiring method may be selected from Chapter 3 based on all the pertinent factors.
>
> **Warning:** You may have wiring methods that are not permitted or have additional restrictions in your jurisdiction. Check local ordinances that amend the NEC®.

If you are not a new student of the NEC®, you may wonder what happened to the requirements for temporary wiring. This article was relocated to Chapter 5 Article 590 (formerly Article 305 and 527)—Temporary Installations. The requirements are essentially still the same as it permits the use of a lower-class wiring for any temporary wiring needed during the construction phase or wiring used for short periods of time such as holiday wiring. These requirements closely parallel OSHA requirements on construction sites. However, OSHA requirements may be much broader, and local experience with OSHA provides guidance in avoiding citations.

The conductor type is selected and sized from Article 310. The wiring method we select may need to be considered at the same time as the conductor type. Most installations require the use of several wiring methods to meet all the conditions, such as a method that is acceptable for dry concealed locations. For this location, most of the methods in Chapter 3 are acceptable. We may also need to select a flexible wiring

method for equipment connection and a material that can be used in wet locations and underground. There are methods that are acceptable in all locations: All are found in Chapter 3, including the junction, outlet, cabinets, and cutout boxes from Article 314 and Article 312. (To identify differences in these two types of enclosures, check Article 100—Definitions.) Once again, I remind you that you are required to comply with all the applicable provisions in Article 300 and the article that covers the specific wiring method.

> **Warning:** Some wiring methods are permitted in locations only when specific type conductors are installed. For example, Flexible Metal Conduit (FMC) (Article 348) is permitted in a wet location when conductors suitable for wet locations are enclosed (310.8(C) and 348.12(1)).

Equipment for General Use

Article 400 and Article 402 cover fixture wires and cords. These articles are used most in connection with equipment and, therefore, are placed in Chapter 4 rather than in Chapter 3.

The requirements for switches (Article 404) and switchboards and panelboards (Article 408) are now covered in Chapter 4. Read the scope of these articles. You see that Article 404 applies to all switches, switching devices, and circuit breakers where used as switches. Article 408's scope is not nearly as clear, and this is where the NEC® can be quite difficult to interpret. Motor control centers are not mentioned in Article 408; a fine print note (FPN No. 1) to NEC® 430.1 states that motor control centers are covered in 110.26(F). You must study the Code carefully and thoroughly to successfully pass your examination the first time you take the test.

Article 404 (formerly Article 380) has been appropriately relocated to Chapter 4 from Chapter 3. It is now easier to find, and it applies to all switches, switching devices, and circuit breakers when they are used as switches.

> **Example:** It is a common practice to use circuit breakers in large commercial installations or warehouses to regularly turn the lights on and off daily. Circuit breakers used as switches must be
>   A. hand operable.
>   B. marked "ON" "OFF."
>   C. marked "SWD" or "HID."
>   D. all of these.

Answer: D. 404.11, they must be of a "hand operable" type; 240.81, they must clearly indicate when they are "ON" and when they are "OFF" and where mounted in a vertical position the "ON" position must be up. *NEC®* 240.83(D): Circuit breakers used as switches in 120-volt and 277-volt fluorescent lighting circuits are required to be listed and marked SWD or HID. Circuit breakers used as switches in high-intensity discharge lighting circuits are required to be listed and marked as HID.

Article 406 covers the rating, type, and installation of receptacles, cord connectors, and attachment plugs (cord caps). These requirements formerly appeared as Part L to Article 410. The requirements are now appropriately located into a separate article and will be much easier for a new student of the *Code* to locate and apply.

Article 408 (formerly Article 384) covers switchboards and panelboards. It has been appropriately relocated to Chapter 4 from Chapter 3. It is now easier to find, and it applies to all switchboards, panelboards, and distribution boards installed for the control of light and power circuits. It also applies to battery-charging panels supplied from light or power circuits.

Article 409—Industrial Control Panels has been added to the *NEC®* for the first time. These industrial control panels are intended for general use and operate from a voltage of 600 volts or less. These control panels are normally manufactured to the UL 508A Standard. They are defined in 409.2 as an assembly of two or more components such as motor controllers, overload relays, fused disconnect switches, and circuit breakers and related control devices but do not include the controlled equipment.

Article 410 covers the requirements for luminaires (lighting fixtures), lampholders, and lamps. The introduction of the term "**luminaire**" was new to the *NEC®* and replaced the term "lighting fixture." The *NEC®* is an international document used in many different countries. Design criteria can be found in this article. *NEC®* 410.4 covers wet locations and gives specific guideline rules for bathrooms where pendant-type luminaires (lighting fixtures) are used. *NEC®* 410.8 covers all rules, allowances, and limitations for luminaires (fixtures) in clothes closets.

Note: High-discharge lighting includes fluorescent luminaires (lighting fixtures) as well as other ballast-type luminaires (lighting fixtures).

As you look through Chapter 4, you will see articles to cover most utilization equipment found in all types of occupancies—that is, appliances, heating, air-conditioning, refrigeration, motors, transformers, phase converters, and generators. Article 490 was added for equipment over 600 volts in the 1999 *NEC®*. It covers the general requirements for equipment operating at more than 600 volts and defines "high voltage" as more than 600 volts. These requirements were formerly found in Article 710, which was deleted.

Special Equipment, Occupancies, and Systems

These chapters amend or modify the first four chapters. Chapter 8 stands alone, and earlier chapters can be applied only where specifically referenced. Chapter 9 contains tables and examples, which are used as applicable. These chapters are addressed in a later chapter of this book.

QUESTION REVIEW

Answer the following Question Review carefully. You will find it necessary to refer to each article in the *National Electrical Code®*, the Table of Contents, and the Index to solve these questions. You will find that answering these questions will prepare you as you proceed through the latter chapters of this book. The next eight chapters of the manual will help you gain that needed understanding of the *NEC®* and give you a wide variety of sample questions on each chapter of the *NEC®*.

> **Each lesson is designed purposely to require the student to apply the entire *NEC®* text, and not specific chapters or articles. It has been found that when studying for a timed, open-book examination, the student must gain proficiency in the Table of Contents, the Index, and the ability to move quickly from cover to cover to find the correct answer to each question in a timely fashion.**

1. Would a floating restaurant located in a river or along a dock in a harbor be covered by the *National Electrical Code®*?

   Answer: _____

   _____

   _____

   _____

   Reference: _____

2. In which article in the *NEC®* would you find the requirements for grounding a community antenna television and radio distribution system?

   Answer: _____

   _____

   _____

   _____

   Reference: _____

3. In which article in the *NEC®* would you find the requirements for receptacles, cord connectors, and attachment plugs?

   Answer: _____

   _____

   _____

   _____

   Reference: _____

4. All requirements for the installation and calculation of branch-circuits are found in Chapter 2 of the *NEC*®. True or False?

Answer: _____

_____

_____

_____

Reference: _____

5. In what article would the requirements for installing luminaires (light fixtures) in a paint spray booth be found?

Answer: _____

_____

_____

_____

Reference: _____

6. In what article would you find the requirements for the installation of central heating equipment, such as a natural gas furnace?

Answer: _____

_____

_____

_____

Reference: _____

7. In what section of the *National Electrical Code*® would you find the allowable number of conductors in a conduit?

Answer: _____

_____

_____

_____

Reference: _____

8. Which article of the *NEC*® covers the power and lumination (lighting) in mine shafts?

Answer: _____

_____

_____

_____

Reference: _____

9. Where would you find the cover requirements for underground conductors running from building A to building B? (The conductors are direct burial conductors, the circuit is a 3-phase, 2300-volt, 4-wire system.)

    Answer: _____

    _____

    _____

    _____

    Reference: _____

10. The conductors in Question 9 must terminate in a disconnecting means. What article and section of the *NEC*®
    cover the requirements for the disconnecting equipment at building B?

    Answer: _____

    _____

    _____

    _____

    Reference: _____

11. A trade size 4 (103) rigid nonmetallic conduit is being installed on the outside of a building. The length of the
    conduit from the junction box to the point that it enters the building is 300 feet (91 m). Where in the *National
    Electrical Code*® is the amount of the expansion of the rigid nonmetallic conduit found?

    Answer: _____

    _____

    _____

    _____

    Reference: _____

12. You are going to wire a nightclub that will seat 300 people. Which article in the NEC® covers the wiring
    requirements for this nightclub?

    Answer: _____

    _____

    _____

    _____

    Reference: _____

13. Where in the *National Electrical Code*® would the requirements for grounding a portable generator be found?

    Answer: _____

    _____

    _____

    _____

    Reference: _____

14. In which section in the *NEC®* would the requirements for bonding the forming shell of an inground swimming pool be found?

Answer: _____

_____

_____

_____

Reference: _____

15. In which section in the *NEC®* is material extracted from another NFPA standard explained?

Answer: _____

_____

_____

_____

Reference: _____

16. Where in the *NEC®* are the requirements for the location and installation of smoke detectors found?

Answer: _____

_____

_____

_____

Reference: _____

17. Under what article of the *NEC®* would the wiring requirements for fire alarms be found?

Answer: _____

_____

_____

_____

Reference: _____

18. An existing building is being remodeled. Permanent receptacles in the building are being used to supply temporary power for the construction. Must these permanent receptacles be GFCI protected?

Answer: _____

_____

_____

_____

Reference: _____

19. You have been given the responsibility for wiring a phosphoric acid fertilizer facility. What section of the *NEC*® covers this type of hazardous installation? (a) What class wiring is required? (b) If the hazardous dusts are not controlled and will be found on the day-to-day work schedule, what division must this area be wired in?

    Answer: _____

    _____

    _____

    _____

    Reference: _____

20. You have been asked to wire a neon sign in the bedroom of a new home you are wiring. To which section of the *NEC*® would you refer to wire this neon sign for the bedroom?

    Answer: _____

    _____

    _____

    _____

    Reference: _____

Chapter Five

OBJECTIVES

Studying this chapter along with the 2005 *NEC®* gives the reader a basic introduction to the general requirements in the *Code* that apply to all installations. The introduction to *NEC®* Chapter 1 covers the general requirements that apply to all installations no matter how small or how large. The reader will study the application of installation rules for above and below 600 volts. The reader will learn how to apply all requirements and understand them.

After completing this chapter, you should know:

- The definitions of words and terms that are unique to the 2005 *NEC®*

- The application of Article 100—Definitions:
  1. Article 100, definitions of words and terms unique to the *NEC®*
  2. Why there are definitions elsewhere in the 2005 *NEC®*
  3. The importance of checking for a term as well as a word

- The application of Article 110—General Requirements

- The definitions of terms such as "listed" or "approved" as they apply to the *NEC®*

- The minimal performance requirements such as "workmanship" and the "application of equipment"

- The application of temperature requirements

- The application of the complex working clearance requirements

GENERAL WIRING REQUIREMENTS

The general requirements for the *National Electrical Code®* are found in Chapter 1. They include Article 100, which is the definitions, Part I—General Definitions, and Part II—the definitions for over 600 volts nominal. Article 110 gives the requirements for electrical installations, Part I—General Requirements, and Part II—the requirements for over 600 volts nominal. Article 100 contains the definitions essential to the proper application of the *National Electrical Code®*. They do not include commonly defined general terms or commonly defined technical terms from other related codes and standards. In general, only the terms used in two or more articles are defined in Article 100. Other definitions are included in the articles for which they are used but may be referenced in Article 100. Part I of Article 100 contains definitions intended to apply wherever the terms are used throughout the *NEC®*. Part II contains only the definitions applicable to installations and equipment operating at over 600 volts nominal. Electrical terms not specific to the *NEC®* may be defined in the *IEEE Dictionary*. Other words are defined in *Webster's Dictionary*. An example of a definition with a specific meaning in the *NEC®* found in Article 100 would be "**accessible**." Accessible is defined both for wiring methods and equipment. The definition found in Article 100 applies specifically to its use throughout the *NEC®*. Other terms or words found in Article 100 have meanings that generally differ from *Webster's Dictionary*.

Example

A bathroom is defined as
- A. a room with a basin, toilet, and tub or a shower.
- B. in a standard dictionary.
- C. an area including a basin, a toilet, and tub or a shower.
- D. an area including a basin with one or more of the following: a toilet, a tub, or a shower.

Answer: D. This definition, found in Article 100, is unique to the *NEC®*. (See Figure 5-1.)

Article 110 contains the general requirements for all electrical installations. It contains the approval required for installations and equipment (see Figure 5–2) and instructions on the examination of the equipment to judge it suitable. It includes information such

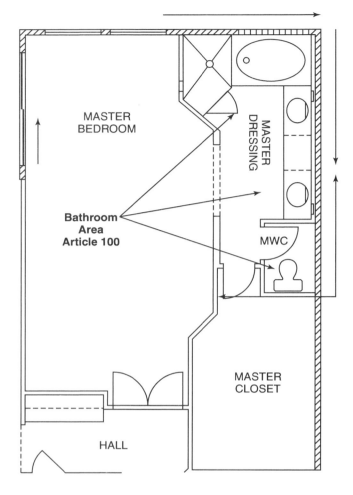

Figure 5–1 This is one example of a "bathroom area." It includes the separate water closet (toilet) room cut off by a door. Another example would be a typical Hotel-Motel "bathroom area" where the basin(s) are located outside the "bathroom" in the sleeping area.

as the voltages considered, conductor information, and many terms to describe the uses throughout the *Code*. *NEC®* 110.8 reminds us that only wiring methods recognized as suitable are included in the *Code*, the recognized methods of wiring permitted to be installed in any type of building or occupancy except as otherwise provided in the *NEC®*. This is important because often materials are misused and wiring methods are devised in the field that are not in accordance with the manufacturer's instructions or design recommendations of the equipment being used. General requirements remind us that all circuitry and equipment intended to break current must be properly sized and installed so proper operation occurs. Article 110 reminds us that the proper equipment must be used

where chemicals, gases, vapors, fumes, liquids, or other agents having a deteriorating effect on equipment must be considered. It reminds us that all work must be installed in a neat and professional manner and that unused openings must be closed.

Wire Temperature Termination Requirements

Wire temperature ratings and temperature termination requirements for equipment result in rejected installations. Information about this topic can be found in testing agency directories, product testing standards, and manufacturers' literature, but most do not consult these sources until it is too late.

In the 2005 *National Electrical Code®*, 110.14(C) provides the information contained in the standards to the inspectors, installers, and engineers.

Why Are Temperature Ratings Important?

Conductors carry a specific temperature rating based on the type of insulation employed on the conductor. Common insulation types can be found in Table 310.13 of the *NEC®*, and corresponding ampacities can be found in Table 310.16.

Example: A 1/0 AWG copper conductor ampacity based on different conductor insulation types:

Insulation Type	Temperature Rating	Ampacity
TW	60°C	125 amperes
THW	75°C	150 amperes
THHN	90°C	170 amperes

Even though the wire size has not changed (1/0 AWG Cu), the ampacity *has* changed due to the temperature rating of the insulation on the conductor. Higher-rated insulation allows a smaller conductor to be used at the same ampacity as a larger conductor with lower-rated insulation, and, as a result, the amount of copper and even the number of conduit runs needed for the job may be reduced.

One common misapplication of conductor temperature ratings occurs when the rating of the equipment termination is ignored. Conductors must be sized by

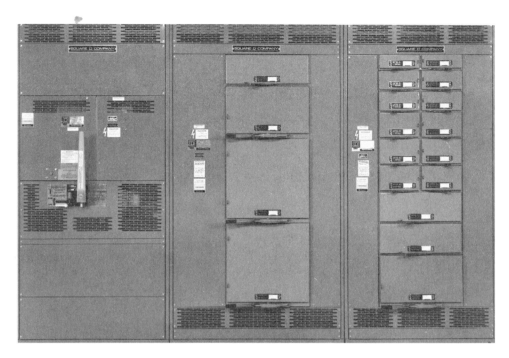

Figure 5–2 Example of an electrical switchboard. *(Courtesy of Square D Company)*

giving consideration to where they will terminate and how that termination is rated. If a termination is rated for 75°C, this means that the temperature at that termination may rise up to 75°C when the equipment is loaded to its ampacity. If 60°C insulated conductors were employed in this example, the additional heat at the connection above the 60°C conductor insulation rating could result in failure of the conductor insulation.

When a conductor is selected to carry a specific load, the user/installer or designer must know termination ratings for the equipment involved in the circuit.

Example: Using a circuit breaker with 75°C termination and a 150-ampere load, if a THHN (90°C) conductor is selected for the job, from Table 310.16 select a conductor that will carry the 150 amperes. Although Type THHN has a 90°C ampacity rating, the ampacity from the 75°C column must be selected because the circuit breaker termination is rated at 75°C. Looking at the table, a 1/0 AWG copper conductor is acceptable. The installation would be as shown in Figure 5–3, with proper heat dissipation at the termination as well as along the conductor length. Had the temperature rating of the termination not been considered, a 1 AWG conductor, based on the 90°C ampacity, may have been selected, which may have led to overheating at the termination or premature opening of the overcurrent device due to the smaller conductor size. (See Figure 5–4.)

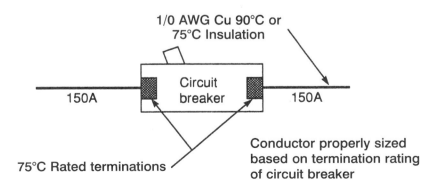

Figure 5–3 *(Courtesy of Square D Company)*

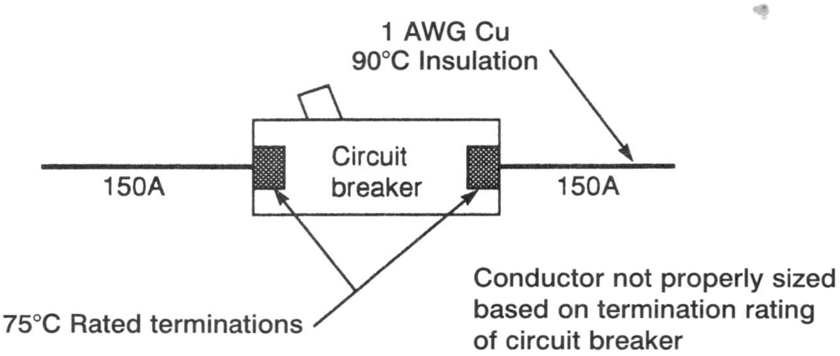

Figure 5–4 *(Courtesy of Square D Company)*

In this example, a conductor with a 75°C insulation type (THW, RHW, USE, etc.) would also be acceptable because the termination is rated at 75°C. A 60°C insulation type (TW or perhaps UF) is not acceptable because the temperature at the termination could rise to a value greater than the insulation rating.

General Rules for Application

When applying equipment with conductor terminations, the following basic rules apply:

- The termination and splicing provisions of equipment shall be based on 110.14(C)(1)(a) or C(1)(b).

- Conductor ampacities used in determining equipment termination provisions shall be based on Table 310.16 as appropriately modified by 310.15 (B)(6).

- ***100.14(C)(1)(a).*** *Termination provisions of equipment for circuits rated **100 amperes or less**, or marked for 14 AWG through 1 AWG, shall be used only for one of the following:*
  (1) Conductors rated 60°C (140°F).
  (2) Conductors with higher temperature ratings, provided the ampacity of such conductors is determined based on the 60°C (140°F) ampacity of the conductor size used.
  (3) Conductors with higher temperature ratings if the equipment is listed and identified for use with such conductors.
  (4) For motors marked with design letters B, C, or D, conductors having an insulation rating of 75°C (167°F) or higher shall be permitted to be used, provided the ampacity of such conductors does not exceed the 75°C (167°F) ampacity.* (See Figure 5-5.)

- ***110.14(C)(1)(b).*** *Termination provisions of equipment for circuits rated **over 100 amperes**, or marked for conductors larger than 1 AWG, shall be used only for one of the following:*
  (1) Conductors rated 75°C (167°F)
  (2) Conductors with higher temperature ratings, provided the ampacity of such conductors does not exceed the 75°C (167°F) ampacity of the conductor size used, or up to their ampacity if the equipment is listed and identified for use with such conductors.* (See Figure 5-6.)

There are exceptions to the rules. Two important exceptions are:

- Conductors with higher temperature insulation may be terminated on lower temperature-rated terminations provided the ampacity of the conductor is based on the lower rating. This is illustrated in the example, where the THHN (90°C) conductor ampacity is based on the 75°C rating to terminate in a 75°C termination. The following table provides a quick reference of how this exception would apply to common terminations.

Conductor Insulation versus Equipment Termination Ratings

Termination Rating	Conductor Insulation Rating		
	60°C	75°C	90°C
60°C	OK	OK (at 60°C ampacity)	OK (at 60°C ampacity)
75°C	No	OK	OK (at 75°C ampacity)
60/75°C	OK	OK (at 60°C or 75°C ampacity)	OK (at 60°C or 75°C ampacity)
90°C	No	No	OK**

** The equipment must have a 90°C rating to terminate 90°C wire at its 90°C ampacity.

* Reprinted with permission from NFPA 70-2005.

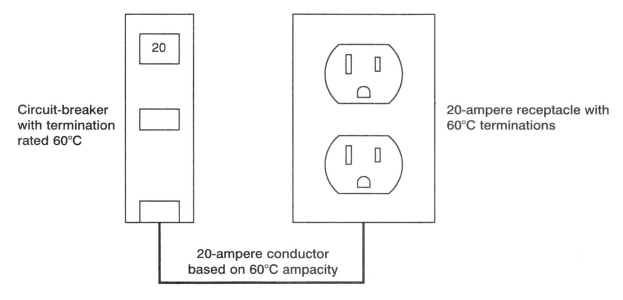

Figure 5–5 *(Courtesy of Square D Company)*

- The termination may be rated for a value higher than the value permitted in the general rules if the equipment is listed and marked for the higher temperature rating.

    Example: A 30-ampere safety switch could have 75°C rated terminations if the equipment was listed and identified for use at this rating.

A Word of Caution

When terminations are inside equipment, such as panelboards, motor control centers, switchboards, enclosed circuit breakers, and safety switches, it is important to note that the temperature rating identified on the equipment labeling should be followed— not the rating of the lug itself. It is common to use 90°C rated lugs (that is, marked AL9CU), but the equipment rating may be only

60°C or 75°C. The use of the 90°C rated lugs in this type of equipment does not give permission for the installer to use 90°C wire at the 90°C ampacity.

The labeling of all devices and equipment should be reviewed for installation guidelines and possible restrictions.

Equipment Terminations Available Today

Remember, a conductor has two ends, and the termination on each end must be considered when applying the sizing rules.

Example: Consider a conductor that will terminate in a 75°C rated termination on a circuit breaker at one end, and a 60°C rated termination on a receptacle at the other end. This circuit must be wired with a conductor that has an insulation rating of at least 75°C (due to the circuit breaker),

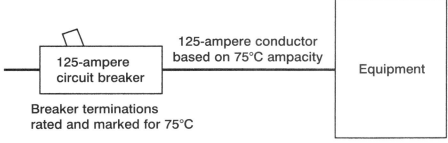

Figure 5–6 *(Courtesy of Square D Company)*

and sized based on the ampacity of 60°C (due to the receptacle).

In electrical equipment, terminations are typically rated at 60°C, 75°C, or 60/75°C. There is no listed distribution or utilization equipment that is listed and identified for the use of 90°C wire at its 90°C ampacity. This includes distribution equipment, wiring devices, transformers, motor control devices, and even utilization equipment, such as HVAC, motors, and luminaires (light fixtures). Installers and designers who have not realized this fact have been faced with jobs that do not comply with the *National Electrical Code*® and jobs that have been turned down by the electrical inspectors.

Example: 90°C wire may be used at its 90°C ampacity (see Figure 5–7). Note that the 2 AWG 90°C rated conductor does not terminate directly in the distribution equipment, but in a terminal or tap box with 90°C rated terminations.

Frequently, manufacturers are asked when distribution equipment will be available with terminations that will permit 90°C conductors at the 90°C ampacity. The answer is complex and requires not only significant equipment redesign (to handle the additional heat),

but also coordination of the downstream equipment where the other end of the conductor will terminate. Significant changes in the product testing/listing standards would also have to occur.

A final note about equipment: Generally, equipment requiring the conductors to be terminated in the equipment has an insulation rating of 90°C but has an ampacity based on 75°C or 60°C. This type of equipment might include 100 percent rated circuit breakers, fluorescent luminaires (lighting fixtures), and so on, and will include a marking to indicate such a requirement. Check with the manufacturer of the equipment to see whether any special considerations need to be taken into account.

Higher-Rated Conductors and Derating Factors

One advantage to conductors with higher insulation ratings is noted when derating factors are applied. Derating factors may be required due to the number of conductors in a conduit, higher ambient temperatures, or possibly internal design requirements for a facility. By beginning the derating process at the ampacity of the conductor based on the higher insulation value, upsizing the conductors to compensate for the derating may not be required.

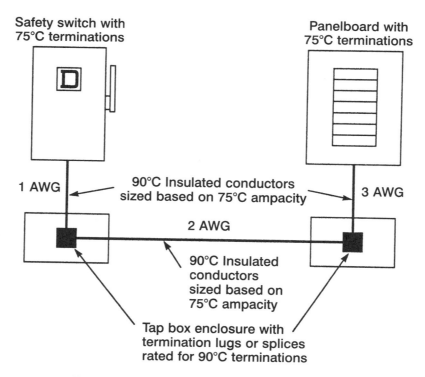

Figure 5–7 *(Courtesy of Square D Company)*

For the following example of this derating process, the following two points must be considered:

1. The ampacity value determined after applying the derating factors must be equal to or less than the ampacity of the conductor, based on the temperature limitations at its terminations.

2. The derated ampacity becomes the allowable ampacity of the conductor, and the conductor must be protected against overcurrent in accordance with this allowable ampacity.

   Example (of the derating process): Assume a 480Y/277 Vac, 3Ø4W feeder circuit to a panelboard supplying 200 amperes of fluorescent luminaire (lighting) load, and assume that the conductors will be in a 40°C ambient temperature. Also, assume that the conductors originate and terminate in equipment with 75°C terminations.

   1. Because the phase and neutral conductors will all be in the same conduit, the issue of conduit fill must be considered. Table 310.15(B)(4) states that the neutral must be considered to be a current-carrying conductor because it is supplying electric discharge lighting.

   2. With four current-carrying conductors in the raceway, apply Table 310.15(B)(2). This requires an 80 percent reduction in the conductor ampacity based on four to six current-carrying conductors in the raceway.

   3. The correction factors at the bottom of Table 310.16 also must be applied. An adjustment of .88 for 75°C and .91 for 90°C is required where applicable.

Now the calculations:

Using a 75°C conductor such as THWN:

300 kcmil copper has a 75°C ampacity of 285 amperes. Using the factors from the example, the calculations are as follows:

$$285 \times .80 \times .88 = 201 \text{ amperes}$$

Thus, 201 amperes is now the allowable ampacity of the 300 kcmil copper conductor for this circuit. Had the derating factors for conduit fill and ambient temperatures not been required, a 3/0 AWG copper conductor would have met these requirements.

Using a 90°C conductor such as THHN:

250 kcmil copper has a 90°C ampacity of 290 amperes. Using the factors from the example, the calculations are as follows:

$$290 \times .80 \times .91 = 211 \text{ amperes}$$

211 amperes is less than the 75°C ampacity of a 250 kcmil copper conductor (255 amperes), so the 211 amperes would now be the allowable ampacity of the 250 kcmil conductor. Had the calculation resulted in a number larger than the 75°C ampacity, the actual 75°C ampacity would have been required to be used as the allowable ampacity of the conductor. This is critical because the terminations are rated at 75°C.

Note: The primary advantage to using 90°C conductors is exemplified here. The conductor is permitted to be reduced by one size (300 kcmil to 250 kcmil) and still accommodate all the required derating factors for the circuit.

In summary, when using 90°C wire for derating purposes, begin by derating at the 90°C ampacity. Compare the result of the calculation to the ampacity of the conductor based on the termination rating (60°C or 75°C). The smaller of the two numbers then becomes the allowable ampacity of the conductor. Note that if the load dictates the size of the required overcurrent device (that is, continuous load × 125%), then the conductor allowable ampacity must be protected by the required overcurrent device. (See 240.4, which permits the conductor to be protected by the next higher standard size overcurrent device as per 240.6.) This may require moving up to a larger conductor size and beginning the derating. In the example, if the 200 amperes of load were continuous, a 250-ampere overcurrent device would be required. The 201 or 211 amperes from our calculations would not be protected by the 250-ampere overcurrent device and, as such, we would be required to move to a larger conductor meeting these requirements to begin the derating process.

Summary

Several factors affect how the allowable ampacity of a conductor is determined. The key is not to treat the wire as a system but as a component of the total electrical system. The terminations, equipment ratings, and environment all affect the ampacity assigned to the conductor. If the designer and the installer remember each rule, the installation will go much more smoothly.

NEC® 110.14 is especially important in that the requirements for all electrical connections must be made in the proper manner. *NEC*® 110.16 reminds us that we must have sufficient working space around electrical equipment of 600 volts nominal or less; that we must have clearances in front and above (refer to

---

110.14 Electrical Connections. Because of different characteristics of dissimilar metals, devices such as pressure terminal or pressure splicing connectors and soldering lugs shall be identified for the material of the conductor and shall be properly installed and used. Conductors of dissimilar metals shall not be intermixed in a terminal or splicing connector where physical contact occurs between dissimilar conductors (such as copper and aluminum, copper and copper-clad aluminum, or aluminum and copper-clad aluminum), unless the device is identified for the purpose and conditions of use. Materials such as solder, fluxes, inhibitors, and compounds, where employed, shall be suitable for the use and shall be of a type that will not adversely affect the conductors, installation, or equipment.

FPN: Many terminations and equipment are marked with a tightening torque.

(A) Terminals. Connection of conductors to terminal parts shall ensure a thoroughly good connection without damaging the conductors and shall be made by means of pressure connectors (including set-screw type), solder lugs, or splices to flexible leads. Connection by means of wire-binding screws or studs and nuts that have upturned lugs or the equivalent shall be permitted for 10 AWG or smaller conductors.

Terminals for more than one conductor and terminals used to connect aluminum shall be so identified.

(B) Splices. Conductors shall be spliced or joined with splicing devices identified for the use or by brazing, welding, or soldering with a fusible metal or alloy. Soldered splices shall first be spliced or joined so as to be mechanically and electrically secure without solder and then be soldered. All splices and joints and the free ends of conductors shall be covered with an insulation equivalent to that of the conductors or with an insulating device identified for the purpose.

Wire connectors or splicing means installed on conductors for direct burial shall be listed for such use.

(C) Temperature Limitations. The temperature rating associated with the ampacity of a conductor shall be selected and coordinated so as not to exceed the lowest temperature rating of any connected termination, conductor, or device. Conductors with temperature ratings higher than specified for terminations shall be permitted to be used for ampacity adjustment, correction, or both.

(1) Equipment Provisions. The determination of termination provisions of equipment shall be based on 110.14 (C)(1)(a) or (C)(1)(b). Unless the equipment is listed and marked otherwise, conductor ampacities used in determining equipment termination provisions shall be based on Table 310.16 as appropriately modified by 310.15(B)(6).

(a) Termination provisions of equipment for circuits rated 100 amperes or less, or marked for 14 AWG through 1 AWG conductors, shall be used only for one of the following:

(1) Conductors rated 60°C (140°F)
(2) Conductors with higher temperature ratings, provided the ampacity of such conductors is determined based on the 60°C (140°F) ampacity of the conductor size used
(3) Conductors with higher temperature ratings if the equipment is listed and identified for use with such conductors
(4) For motors marked with design letters B, C, D, or E, conductors having an insulation rating of 75°C (167°F) or higher shall be permitted to be used provided the ampacity of such conductors does not exceed the 75°C (167°F) ampacity.

(b) Termination provisions of equipment for circuits rated over 100 amperes, or marked for conductors larger than 1 AWG, shall be used only for one of the following:

(1) Conductors rated 75°C (167°F)
(2) Conductors with higher temperature ratings, provided the ampacity of such conductors does not exceed the 75°C (167°F) ampacity of the conductor size used, or up to their ampacity if the equipment is listed and identified for use with such conductors.

(2) Separate Connector Provisions. Separately installed pressure connectors shall be used with conductors at the ampacities not exceeding the ampacity at the listed and identified temperature rating of the connector.

FPN: With respect to 110.14(C)(1) and (2), equipment markings or listing information may additionally restrict the sizing and temperature ratings of connected conductors.

Figure 5–2); we must have headroom to work on the equipment; proper illumination; and that all live parts must be guarded against accidental contact. Article 110 reminds us that warning signs must be placed and that electrical equipment must be protected from physical damage. *NEC®* 110.22 states that each disconnecting means for motors and appliances, and service, feeders, or branch circuits at the point where it originates must be legibly identified to indicate its purpose and that the identifying markings must be sufficiently durable to withstand the environment involved. Where circuits and fuses are installed with a series combination rating "Caution—Series Rated System," the equipment enclosure shall be legibly marked in the field to indicate that the equipment has been applied with a series combination rating. Over-600-volt requirements are found in Part III of Article 110. These requirements are specific about entrance and access to working space, guarding barriers, enclosures for electrical installations (refer to Figure 5–2), and separation from other circuitry. Over-600-volt installations must be located in locked rooms or enclosures, except where under the observation of qualified persons at all times. Table 110.34(E) covers the elevation of unguarded live parts above working space for nominal voltages 601 through over 35 kV.

Working Clearances

NEC® 110.26 reminds us that we must have sufficient working space around electrical equipment of 600 volts nominal or less; that we must have clearances in front and above (see switchboard, Figure 5–2), and that all live parts must be guarded against accidental contact. Article 110.26 reminds us that warning signs must be placed and that electrical equipment must be protected from physical damage. (See Figure 5–8.)

NEC® 110.26(E) states that a minimum headroom of 6½ feet (2 m) or to the top of the equipment, whichever is greater, must be maintained in the required working spaces to work on the equipment. This means that the area 30 inches (750 mm) wide or the width of the equipment, whichever is greater, by the depth required in Table 110.26(A)(1) must have a minimum headroom clearance of the height of the equipment or 6½ feet (2 m), whichever is greater, to provide a safe place for the worker to work on the equipment. Proper illumination must be provided in these working spaces and around the equipment. If an adequate lighting source is provided in these rooms or areas, additional specified lighting is not required.

NEC® 110.22 states that each disconnecting means for motors and appliances, and service, feeders, or branch circuits at the point where it originates must be legibly identified to indicate its purpose. Also, the identifying markings must be sufficiently durable to withstand the environment involved. Where circuits and fuses are installed with a series combination rating "Caution—Series Rated System," the equipment enclosure shall be legibly marked in the field to indicate that the equipment has been applied with a series combination rating.

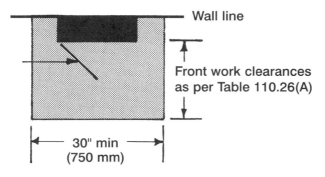

Top view of electrical equipment

Wall line

Front work clearances as per Table 110.26(A)

30" min (750 mm)

In all cases the door must be able to open 90°.

Figure 5–8 *NEC®* 110.26(A) requires that a minimum of clearance of 30 inches (750 mm) wide be maintained in front of electrical equipment or the width of the equipment, whichever is greater; in addition, all doors must have clearance to open at least 90°. The distance in front of the equipment must be in accordance with Table 110.26(A). See Part III to Article 110 for required clearances over 600 volts and Table 110.34(A) for clearances.

Working Clearances Over 600 Volts

Over-600-volt requirements are found in Part III, which supplements or modifies the preceding sections. In no case shall the provisions of this part apply to equipment on the supply-side of the service conductors. These requirements are very specific regarding entrance and access to working space, guarding barriers, enclosures for electrical installations, and separation from other circuitry (see Figure 5–8 and Figure 5–9).

Over-600-volt installations must be located in locked rooms or enclosures, except where under the observation of qualified persons at all times. Table 110.34(E) covers the elevation of unguarded live parts above working space for nominal voltages 601 through over 35 kV.

Many of the requirements for over-600-volt installations are similar to those for 600 volts or less; however, when preparing for an examination or applying these sections, it is very important that these sections be studied carefully.

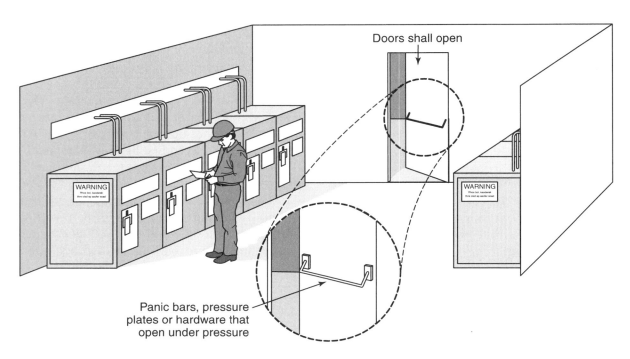

Figure 5–9 *NEC®* 110.26(C) (up to 600 volts) and 110.33(A)(1) (over 600 volts) defines large equipment and the required entrance and access to large electrical equipment. Where the qualified worker is required to exit through a door to get out of the working space, the door shall open out and be equipped with panic-type hardware.

QUESTION REVIEW

> Each lesson is designed purposely to require the student to apply the entire *NEC®* text, and not specific chapters or articles. It has been found that when studying for a timed, open-book examination, the student must gain proficiency in the Table of Contents, the Index, and the ability to move quickly from cover to cover to find the correct answer to each question in a timely fashion.

The following drawing applies to Questions 1–5.

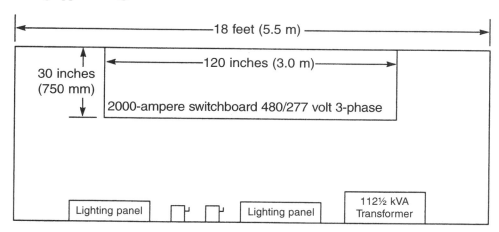

This proposed electrical equipment room is to be constructed of concrete floor and concrete. Block walls will be painted.

1. What is the space required between the switchboard and the 112½ kVA transformer?

   Answer: _____

   _____

   _____

   _____

   Reference: _____

2. How many doors are required for entering and leaving this room?

   Answer: _____

   _____

   _____

   _____

   Reference: _____

3. Is this electrical equipment room required to be illuminated?

   Answer: _____

   _____

   _____

   _____

   Reference: _____

4. What is the required headroom in front of the electrical equipment needed for those performing tests or maintenance on this electrical equipment?

    Answer: _____

    _____

    _____

    _____

    Reference: _____

5. If the proposed electrical room size was increased from 10×18 to 18×18, how many doors would be required, provided that the room had no additional electrical equipment added?

    Answer: _____

    _____

    _____

    _____

    Reference: _____

6. Does the heat pump installation shown here comply with the *NEC*®? This is a single-family dwelling. The supplementary overcurrent device and required disconnecting means are located on the dwelling wall directly behind the heat pump for convenience.

    Answer: _____

    _____

    _____

    _____

    Reference: _____

7. The working space in front of enclosed electrical equipment operating at 480 volts must have a width when facing the equipment of at least _____ inches (_____ mm).

   Answer: _____

   _____

   _____

   _____

   Reference: _____

8. In an electrical system, an overcurrent device shall be placed in series with each _____ conductor.

   Answer: _____

   _____

   _____

   _____

   Reference: _____

9. A conduit nipple of 18 inches (450 mm) in length is installed between two boxes. This nipple can be filled to a maximum of _____ percent without derating the conductors.

   Answer: _____

   _____

   _____

   _____

   Reference: _____

10. A distribution panelboard must *not* be installed in a _____.

    Answer: _____

    _____

    _____

    _____

    Reference: _____

11. An unprotected cable is installed through bored holes in wood studs. The holes must be bored so that the edge of the hole is at least _____ inch(es) (_____ mm) from the nearest edge of the stud.

    Answer: _____

    _____

    _____

    _____

    Reference: _____

12. Circuit conductors run in electrical nonmetallic tubing (ENT) cannot exceed _____ volts.

    Answer: _____

    _____

    _____

    _____

    Reference: _____

13. Can liquidtight flexible nonmetallic conduit be used as a service raceway?

    Answer: _____

    _____

    _____

    _____

    Reference: _____

14. When a metal raceway is used as physical protection for a grounding electrode conductor, must this metallic raceway be bonded even though the grounding electrode conductor is bare copper? If the answer is yes, how must this metal raceway be bonded?

    Answer: _____

    _____

    _____

    _____

    Reference: _____

15. The size of the equipment grounding conductor routed with the feeder or branch-circuit conductors is determined by what?

    Answer: _____

    _____

    _____

    _____

    Reference: _____

16. How many outdoor receptacle outlets are required for a one-family dwelling?

    Answer: _____

    _____

    _____

    _____

    Reference: _____

17. Are the metal parts of electrical equipment associated with a hydromassage tub required to be bonded?

    Answer: _____

    _____

    _____

    _____

    Reference: _____

18. Where rooms within dwellings are separated by railings, planters, and so forth, are receptacles required to be installed in or along these items?

    Answer: _____

    _____

    _____

    _____

    Reference: _____

19. Are the receptacle outlet spacings for room dividers, bar-type counters, and fixed room dividers the same for site-built dwellings as they are for mobile homes?

    Answer: _____

    _____

    _____

    _____

    Reference: _____

20. You have been asked to install a receptacle on a rooftop for the air-conditioning repair people to plug their instruments and portable tools into. In which section of the *NEC®* would this requirement be found?

    Answer: _____

    _____

    _____

    _____

    Reference: _____

Chapter Six

BRANCH-CIRCUITS AND FEEDERS

Branch-circuits are covered in *NEC®* Article 210. Part I covers the general provisions for branch-circuits; Part II covers the branch-circuit ratings; and Part III gives the required outlets. Outdoor branch-circuits and feeders are found in *NEC®* Article 225. Article 215 covers other feeders, and *NEC®* Article 220 covers the calculations for branch-circuits and feeders and the optional calculations for calculating feeder and service loads. However, branch-circuits are found throughout the *Code*. Air-conditioning branch-circuits are found in *NEC®* Article 440, appliance branch-circuits in

Branch Circuit
The circuit conductors between the final over-current device protecting the circuit and the outlet(s).

Branch Circuit, Appliance
A branch circuit that supplies energy to one or more outlets to which appliances are to be connected and that has no permanently connected luminaires (lighting fixtures) that are not a part of an appliance.

Branch Circuit, General-Purpose
A branch circuit that supplies two or more receptacles or outlets for lighting and appliances.

Branch Circuit, Individual
A branch circuit that supplies only one utilization equipment.

Branch Circuit, Multiwire
A branch circuit that consists of two or more un-grounded conductors that have a voltage between them, and a grounded conductor that has equal voltage between it and each ungrounded conductor of the circuit and that is connected to the neutral or grounded conductor of the system.

Feeder
All circuit conductors between the service equipment, the source of a separately derived system, or other power supply source and the final branch-circuit overcurrent device.

NEC® Article 422, heating branch-circuits in *NEC®* Article 424, and motor branch-circuits in *NEC®* Article 430, and so forth. Chapters 5, 6, and 7 supplement the first four chapters of the *Code*. Therefore, for branch-circuit studies, it is necessary that you be familiar with the entire *NEC®* and utilize the Index to quickly find the specific type branch-circuits with which you are dealing.

Branch-circuits are defined in *NEC®* Article 100 and are defined as a branch-circuit, an appliance branch-circuit, a general-purpose branch-circuit, an individual branch-circuit, a multiwire branch-circuit, and the branch-circuit selection current. It is very important that you read these definitions before applying the requirements in the *NEC*, so that you are properly applying the requirements.

A feeder is defined in Article 100 as the circuit conductor between the service equipment or the source of a separately derived system and the final branch-circuit overcurrent device.

Caution: A feeder generally terminates in more than one overcurrent device.

For instance, many luminaires (lighting fixtures) and other equipment contain supplementary overcurrent devices at the fixture or equipment. The conductors feeding that equipment should still be considered as the branch-circuit conductors and not feeder conductors. Feeders are generally found in *NEC* Article 215 and Article 225. Calculations for sizing feeders are found in *NEC* Article 220. Other references to feeders can be found in *NEC* Article 368 for busways, *NEC* Article 430 for motors, *NEC* Article 550 for

mobile homes, and *NEC* 530 for motion picture studios. See the Index for a complete listing of where feeders can be found in the *NEC*. (See Figure 6–1.)

General-Purpose Branch-Circuits

The simplest forms of branch-circuits are general-purpose branch-circuits, which supply a number of outlets for luminaires (lighting) and appliances, and these are in 15-, 20-, 30-, 40-, and 50-ampere sizes. (See Figure 6–2.) Where conductors or higher ampacity are used for any reason, the ampere rating or the setting of the specified overcurrent device determines the circuit classification. Multiwire branch-circuits greater than 50 amperes can be permitted for nonlighting outlet loads on industrial premises where maintenance and supervision indicate that a qualified person will service the equipment. A common multioutlet branch-circuit greater than 50 amperes is often found in industrial buildings utilizing welding receptacles so that the maintenance personnel can move their welders

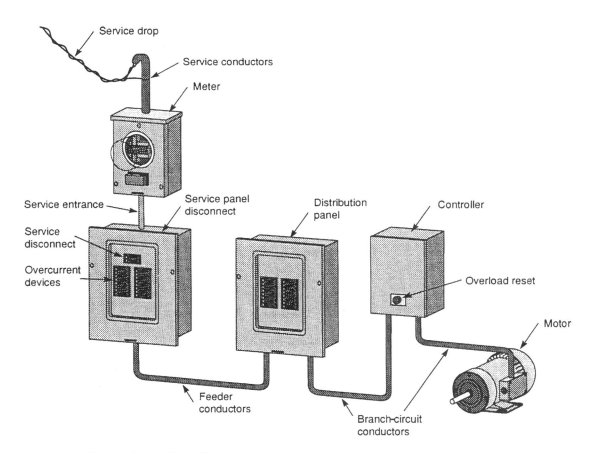

Figure 6–1 Specific conductor types as defined in *NEC* Article 100.

Example D6 Maximum Demand for Range Loads

Table 220.55, Column C applies to ranges not over 12 kW. The application of Note 1 to ranges over 12 kW (and not over 27 kW) and Note 2 to ranges over 8¾ kW (and not over 27 kW) is illustrated in the following two examples.

A. Ranges All the Same Rating *(see Table 220.55, Note 1)*
Assume 24 ranges, each rated 16 kW.

From Table 220.55 Column C, the maximum demand for 24 ranges of 12-kW rating is 39 kW. 16 kW exceeds 12 kW by 4.

5% × 4 = 20% (5% increase for each kW in excess of 12)
39 kW × 20% = 7.8 kW increase
39 kW + 7.8 kW = 46.8 kW (value to be used in selection of feeders)

B. Ranges of Unequal Rating *(see Table 220.55, Note 2)*
Assume 5 ranges, each rated 11 kW; 2 ranges, each rated 12 kW; 20 ranges, each rated 13.5 kW; 3 ranges, each rated 18 kW.

5 ranges × 12 kW	=	60 kW (use 12 kW for range rated less than 12)
2 ranges × 12 kW	=	24 kW
20 ranges × 13.5 kW	=	270 kW
3 ranges × 18 kW	=	54 kW
30 ranges Total kW	=	408 kW

408 kW ÷ 30 ranges = 13.6 kW (average to be used for calculation)

From Table 220.55 Column C, the demand for 30 ranges of 12-kW rating is 15 kW + 30 (1 kW × 30 ranges) = 45 kW. 13.6 kW exceeds 12 kW by 1.6 kW (use 2 kW).

5% × 2 = 10% (5% increase for each kW in excess of 12 kW)

45 kW × 10% = 4.5 kW increase

45 kW + 4.5 kW = 49.5 kW (value to be used in selection of feeders)

Figure 6–2 Branch-circuit, feeder, and service-load calculations for a single-family dwelling. (Reprinted with permission from NFPA 70-2005, the *National Electrical Code®*, Copyright © 2004–, National Fire Protection Association, Quincy, MA 02269. This reprinted material is not the complete and official position of the National Fire Protection Association on the referenced subject, which is represented only by the standard in its entirety.)

throughout the plant on an as-needed basis. In these instances, you may find many receptacles on one 60-ampere or above rated circuit to supply these receptacles. Generally, no danger of overloading this circuit exists because the plant has a limited number of welding machines and personnel capable of utilizing those machines. There are no voltage limitations on branch-circuits, because branch-circuits can supply the equipment and motors of many varying different voltages.

Example: What is the maximum circuit rating for a cord-and-plug-connected load that can be connected to a circuit that contains two or more 15-ampere receptacles? What is the maximum load that may be connected?

Step 1. Refer to the Index "Branch Circuits, Overcurrent Protection," which references 210.20.

Step 2. *NEC®* 210.20 gives the rules for 15- and 20-ampere branch-circuits and references Table 210.21(B)(2).

Step 3. Table 210.21(B)(2).

Answer: A 15- or 20-ampere circuit with a maximum connected load of 12 amperes.

Example: How many general-use, duplex 20-ampere receptacles can be connected to one 120-volt, 20-ampere circuit-breaker?

Step 1. This question is not complete enough to reach a single answer; therefore, we must look at both residential dwelling units and at other type installations.

Step 2. Refer to the Table of Contents; branch-circuit calculations are found in Article 220.

Step 3. Calculation of branch-circuits, 220.10.

Step 4. Table 220.12; from the table we see "Dwelling Units." See 220.14(J).

Step 5. **Dwelling Occupancies.** *In one-family, two-family, and multifamily dwellings and in guest rooms or guest suites of hotels and motels, the outlets specified in J(1), J(2), and J(3) are included in the general lighting load calculations of 220.12. No additional load calculations shall be required for such outlets.*
(1) All general-use receptacle outlets of 20-ampere rating or less, including receptacles connected to the circuits in 210.11(C)(3)*

* Reprinted with permission from NFPA 70-2005.

(2) The receptacle outlets specified in 210.52(E) and (G)

(3) The lighting outlets specified in 210.70(A) and (B)*

Answer: For dwelling units, the number is not limited.

Step 6. For all other occupancies we must go to 220.16(B) because we have no special instructions under Table 220.12.

Step 7. **Other Outlets.** Other outlets not covered shall be calculated based on 220.14(I) 180 volt-amperes per outlet.

Step 8. I = VA ÷ E
I = 180 ÷ 120
I = 1.5 amperes.

Step 9. Table 210.24 references 210.23(A), which limits cord-and-plug-connected loads to 80 percent of the branch-circuit rating.

Step 10. 20-ampere circuit-breaker multiplied by 80 percent equals 16 amperes.

Step 11. 16 ÷ 1.5 = 10.667.

Answer: For other occupancies, ten duplex receptacles are permitted.

NEC® 210.6 lists the limitations for branch-circuit voltages. In occupancies such as dwelling units and guest rooms of hotels, motels, and similar occupancies, the voltage is not permitted to exceed 120 volts between conductors that supply terminals of luminaires (lighting fixtures) and cord-and-plug-connected loads 1440 VA nominal or less, or less than ¼ horsepower, and permitted to supply terminals of medium-base, screw-shell lampholders, or lampholders of other types applied within their voltage ratings, auxiliary equipment of electric discharge lamps, and cord-and-plug-connected or permanently connected utilization equipment. Circuits exceeding 120 volts, nominal, and not exceeding 277 volts, nominal, to ground are permitted to supply listed electric-discharge luminaires (lighting fixtures) equipped with medium-base screw-shell lampholder, luminaires (lighting fixtures) with mogul-base screw-shell lampholders, and lampholders other than the screw-shell type applied within their voltage ratings, auxiliary equipment of electric discharge lamps, and cord-and-plug connected or permanently connected utilization equipment. Circuits exceeding 277 volts, nominal, to ground and not exceeding 600

volts, nominal, between conductors are permitted to supply auxiliary equipment of electric discharge lamps mounted in permanently installed luminaires (fixtures) where the luminaires (fixtures) are mounted in accordance with *NEC®* 210.6(D) (see Figure 6–3.) Branch-circuits supplying equipment over 600 volts are required to comply with the appropriate *Code* sections and articles related to that equipment.

> **Example:**
> The circuit conductors supplying a 600-horse-power, 2300-volt, 3-phase induction motor are described in the 2005 *NEC®* as a
>   A. feeder.
>   B. branch circuit.
>   C. service conductor.
>   D. none of the above.
>
> **Answer:** B. **Branch circuit.** *The circuit conductors between the final overcurrent device protecting the circuit and the outlet(s)**
>

Ground-Fault Protection

Ground-fault circuit-interrupter protection for personnel for dwelling units and other location requirements can be found in *NEC®* 210.8. However, other requirements for ground-fault circuit interrupters can be found in specific articles, such as Article 511 for commercial garages, Article 517 for health care facilities, and Article 680 for swimming pools. *NEC®* 210.52 covers the requirements and general provisions for installing receptacle outlets in dwellings and does not cover installations other than dwelling installations generally; there are some exceptions. Provisions are in the *Code*, in Article 220, that do cover these receptacle outlets in other than dwelling installations when installed. Lighting outlets for dwellings are covered in *NEC®* 210.70. Just as for receptacle outlets, the lighting requirements are generally not specified in the *NEC®* other than for dwellings. However, the installation procedures are specified when they are installed in these installations. When calculating the loads for branch-circuit conductors, one must verify whether these loads are rated as continuous duty or as noncontinuous duty. Continuous duty is defined in Article 100. Most loads in dwellings are considered to be noncontinuous loads because they would be on for less than three hours at any given time. An example of a continuous load would be the lighting in a commercial

Branch-circuits supplying lampholders—Exceptions to 150 volt limit

Incandescent lamps

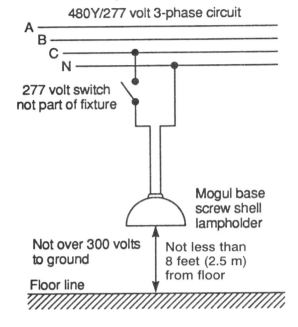

Electric discharge lamps

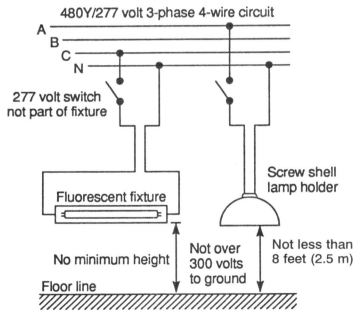

Figure 6–3 Examples of incandescent and electric discharge luminaires (lighting fixtures). *(Courtesy of American Iron & Steel Institute)*

or industrial establishment, which would generally be energized for more than three hours at a time. Most branch-circuit calculations are made in Article 220. (See Figure 6–4.) However, specific branch-circuit requirements for branch-circuits supplying specific utilization equipment are also found in the appropriate article for that equipment, such as Article 440 for air-conditioning equipment and Article 430 for motors. (See Figure 6–5, page 89.) There are also many other

(Text continues on page 90)

Example D1(a). One-Family Dwelling

The dwelling has a floor area of 1500 ft^2, exclusive of an unfinished cellar not adaptable for future use, unfinished attic, and open porches. Appliances are a 12-kW range and a 5.5-kW, 240-V dryer. Assume range and dryer kW ratings equivalent to kVA ratings in accordance with 220.54 and 220.55.

Computed Load *[see 220.40]*
General Lighting Load: 1500 ft^2 at 3 VA per ft^2 = 4500 VA

Minimum Number of Branch Circuits Required *[see 210.11(A)]*
General Lighting Load: 4500 VA ÷ 120 V = 37.5 A

This requires three 15-A, 2-wire or two 20-A, 2-wire circuits
Small Appliance Load: Two 2-wire, 20-A circuits *[see 210.11(C)(1)]*
Laundry Load: One 2-wire, 20-A circuit *[see 210.11(C)(2)]*
Bathroom Branch Circuit: One 2-wire, 20-A circuit (no additional load calculation is required for this circuit) *[see 210.11(C)(3)]*

Minimum Size Feeder Required *[see 220.40]*

General Lighting		4500 VA
Small Appliance		3000 VA
Laundry		1500 VA
	Total	9000 VA
3000 VA at 100%		3000 VA
9000 VA − 3000 VA =		
6000 VA at 35%		2100 VA
	Net Load	5100 VA
Range *(see Table 220.19)*		8000 VA
Dryer *(see Table 220.54)*		5500 VA
	Net Calculated Load	18,600 VA

Net Computed Load for 120/240-V, 3-wire, single-phase service or feeder

$$18{,}600 \text{ VA} \div 240 \text{ V} = 77.5 \text{ A}$$

Sections 230.42(B) and 230.79 require service conductors and disconnecting means rated not less than 100 amperes.

Calculation for Neutral for Feeder and Service

Lighting and Small Appliance Load		5100 VA
Range: 8000 VA at 70% *(see 220.61)*		5600 VA
Dryer: 5500 VA at 70% *(see 220.61)*		3850 VA
	Total	14,550 VA

Calculated Load for Neutral

$$14{,}550 \text{ VA} \div 240 \text{ V} = 60.6 \text{ A}$$

Example D1(b). One-Family Dwelling

Assume same conditions as Example No. D1(a), plus addition of one 6-A, 230-V, room air-conditioning unit and one 12-A, 115-V, room air-conditioning unit,* one 8-A, 115-V rated waste disposer, and one 10-A, 120-V rated dishwasher.* See Article 430 for general motors and Article 440, Part VII, for air-conditioning equipment. Motors have nameplate ratings of 115 V and 230 V for use on 120-V and 240-V nominal voltage systems.

* (For feeder neutral, use larger of the two appliances for unbalance.)

From Example No. D1(a), feeder current is 78 A (3-wire, 240 V).

	Line A	Neutral	Line B
Amperes from Example D1(a)	78	61	78
One 230-V air conditioner	6	—	6
One 115-V air conditioner and 120-V dishwasher	12	12	10
One 115-V disposer	—	8	8
25% of largest motor *(see 430.24)*	3	3	2
Total amperes per line	99	84	104

Therefore, the service would be rated 110 A.

Example D2(a). Optional Calculation for One-Family Dwelling, Heating Larger than Air-Conditioning *[see 220.82]*

The dwelling has a floor area of 1500 ft^2, exclusive of an unfinished cellar not adaptable for future use, unfinished attic, and open porches. It has a 12-kW range, a 2.5-kW water heater, a 1.2-kW dishwasher, 9 kW of electric space heating installed in five rooms, a 5-kW clothes dryer, and a 6-A, 230-V, room air-conditioning unit. Assume range, water heater, dishwasher, space heating, and clothes dryer kW ratings equivalent to kVA.

Air Conditioner kVA Calculation

$$6 \text{ A} \times 230 \text{ V} \div 1000 = 1.38 \text{ kVA}$$

This 1.38 kVA [Item 1 from 220.82(C)] is less than 40% of

Figure 6–4 Examples of branch-circuit, feeder, and service-load calculations for different types of occupancies. (Reprinted with permission from NFPA 70-2005, the *National Electrical Code®*, Copyright © 2004–, National Fire Protection Association, Quincy, MA 02269. This reprinted material is not the complete and official position of the National Fire Protection Association on the referenced subject, which is represented only by the standard in its entirety.)

9 kVA of separately controlled electric heat [Item 6 from 220.82(C)], so the 1.38 kVA need not be included in the service calculation.

General Load

1500 ft² at 3 VA	4500 VA
Two 20-A appliance outlet circuits at 1500 VA each	3000 VA
Laundry circuit	1500 VA
Range (at nameplate rating)	12,000 VA
Water heater	2500 VA
Dishwasher	1200 VA
Clothes dryer	5000 VA
Total	29,700 VA

Application of Demand Factor [See 220.82(B)]

First 10 kVA of general load at 100%	10,000 VA
Remainder of general load at 40% (19.7 kVA × 0.4)	7880 VA
Total of general load	17,880 VA
9 kVA of heat at 40% (9000 VA × 0.4) =	3600 VA
Total load	21,480 VA

Calculated Load for Service Size

$$21.48 \text{ kVA} = 21,480 \text{ VA}$$

$$21,480 \text{ VA} \div 240 \text{ V} = 89.5 \text{ A}$$

Therefore, the minimum service size would be 100 A in accordance with 230.42 and 230.79.

Feeder Neutral Load, per 220.61

1500 ft² at 3 VA		4500 VA
Three 20-A circuits at 1500 VA		4500 VA
	Total	9000 VA
3000 VA at 100%		3000 VA
9000 VA − 3000 VA = 6000 VA at 35%		2100 VA
	Subtotal	5100 VA
Range: 8 kVA at 70%		5600 VA
Clothes dryer: 5 kVA at 70%		3500 VA
Dishwasher		1200 VA
	Total	15,400 VA

Calculated Load for Neutral

$$15,400 \text{ VA} \div 240 \text{ V} = 64.2 \text{ A}$$

Example D2(b). Optional Calculation for One-Family Dwelling, Air Conditioning Larger than Heating [see 220.82(A) and 220.82(C)]

The dwelling has a floor area of 1500 ft², exclusive of an unfinished cellar not adaptable for future use, unfinished attic, and open porches. It has two 20-A small appliance circuits, one 20-A laundry circuit, two 4-kW wall-mounted ovens, one 5.1-kW counter-mounted cooking unit, a 4.5-kW water heater, a 1.2-kW dishwasher, a 5-kW combination clothes washer and dryer, six 7-A, 230-V room air-conditioning units, and a 1.5-kW permanently installed bathroom space heater. Assume wall-mounted ovens, counter-mounted cooking unit, water heater, dishwasher, and combination clothes washer and dryer kW ratings equivalent to kVA.

Air Conditioning kVA Calculation

$$\text{Total amperes} = 6 \text{ units} \times 7 \text{ A} = 42 \text{ A}$$

$$42 \text{ A} \times 240 \text{ V} \div 1000 = 10.08 \text{ kVA (assume PF} = 1.0)$$

Load Included at 100%
Air-Conditioning: Included below [see item 1 in 220.82(C)]
Space Heater: Omit [see item 5 in 220.82(C)]

General Load

1500 ft² at 3 VA	4500 VA
Two 20-A small appliance circuits a at 1500 VA each	3000 VA
Laundry circuit	1500 VA
Two ovens	8000 VA
One cooking unit	5100 VA
Water heater	4500 VA
Dishwasher	1200 VA
Washer/dryer	5000 VA
Total general load	32,800 VA
First 10 kVA at 100%	10,000 VA
Remainder at 40% (22.8 kVA × 0.4 × 1000)	9120 VA
Subtotal general load	19,120 VA
Air conditioning	10,080 VA
Total	29,200 VA

Calculated Load for Service

$$29,200 \text{ VA} \div 240 \text{ V} = 122 \text{ A (service rating)}$$

Feeder Neutral Load, per 220.61
Assume that the two 4-kVA wall-mounted ovens are supplied

Figure 6–4 *(continued)* (Reprinted with permission from NFPA 70-2005, the *National Electrical Code*®, Copyright © 2004–, National Fire Protection Association, Quincy, MA 02269. This reprinted material is not the complete and official position of the National Fire Protection Association on the referenced subject, which is represented only by the standard in its entirety.)

by one branch-circuit, the 5.1-kVA counter-mounted cooking unit by a separate circuit.

1500 ft² at 3 VA		4500 VA
Three 20-A circuits at 1500 VA		4500 VA
	Subtotal	9000 VA
3000 VA at 100%		3000 VA
9000 VA – 3000 VA = 6000 VA		
at 35%		2100 VA
	Subtotal	5100 VA

Two 4-kVA ovens plus one 5.1-kVA cooking unit = 13.1 kVA. Table 220.55 permits 55% demand factor or, 13.1 kVA × 0.55 = 7.2 kVA feeder capacity.

	Subtotal from above	5100 VA
Ovens and cooking unit:		
7200 VA × 70% for neutral load		5040 VA
Clothes washer/dryer:		
5 kVA × 70% for neutral load		3500 VA
Dishwasher		1200 VA
	Total	14,840 VA

Calculated Load for Neutral

14,840 VA ÷ 240 V = 61.83 A (use 62 A)

Example D2(c). Optional Calculation for One-Family Dwelling with Heat Pump (Single-Phase, 240/120-Volt Service) *(see 220.82)*

The dwelling has a floor area of 2000 ft², exclusive of an unfinished cellar not adaptable for future use, unfinished attic, and open porches. It has a 12-kW range, a 4.5-kW water heater, a 1.2-kW dishwasher, a 5-kW clothes dryer, and a 2½-ton (24-A) heat pump with 15 kW of backup heat.

Heat Pump kVA Calculation

24 A × 240 V ÷ 1000 = 5.76 kVA

This 5.76 kVA is less than 15 kVA of the backup heat; therefore, the heat pump load need not be included in the service calculation *[see 220.82(C)]*.

General Load

2000 ft² at 3 VA	6000 VA
Two 20-A appliance outlet circuits	
at 1500 VA each	3000 VA
Laundry circuit	1500 VA
Range (at nameplate rating)	12,000 VA

Water heater	4500 VA
Dishwasher	1200 VA
Clothes dryer	5000 VA
Subtotal general load	33,200 VA
First 10 kVA of general load at 100%	10,000 VA
Remainder of general load at 40%	
(23,200 VA × 0.4)	9280 VA
Total net general load	19,280 VA

Heat Pump and Supplementary Heat*

240 V × 24 A = 5760 VA

15-kW Electric Heat:

5760 VA + (15,000 VA × 65%) = 5.76 kVA + 9.75 kVA = 15.51 kVA

*If supplementary heat is not on at same time as heat pump, heat pump kVA need not be added to total.

Totals

Net general load		19,280 VA
Heat pump and supplementary heat		15,510 VA
	Total	34,790 VA

Calculated Load for Service

34.79 kVA × 1000 ÷ 240 V = 144.96 A

This dwelling unit would be permitted to be served by a 150-A service.

Example D3. Store Building

A store 50 ft by 60 ft, or 3000 ft², has 30 ft of show window. There are a total of 80 duplex receptacles. The service is a 120/240 V, single-phase 3-wire service. Actual connected lighting load is 8500 VA.

Computed Load *(see 220.40)*

Noncontinuous Loads
Receptacle Load *(see 220.44)*

80 receptacles at 180 VA		14,400 VA
10,000 VA at 100%		10,000 VA
14,400 VA – 10,000 VA =		
4400 VA at 50%		2,200 VA
	Subtotal	12,200 VA

Figure 6–4 *(continued)* (Reprinted with permission from NFPA 70-2005, the *National Electrical Code®*, Copyright © 2004–, National Fire Protection Association, Quincy, MA 02269. This reprinted material is not the complete and official position of the National Fire Protection Association on the referenced subject, which is represented only by the standard in its entirety.)

Continuous Loads

General Lighting*

3000 ft² at 3 VA per ft²	9000 VA
Show Window Lighting Load	
30 ft at 200 VA per ft	6000 VA
Outside Sign Circuit *[see 220.14(F)]*	1200 VA
Subtotal	16,200 VA

Subtotal from noncontinuous	12,200 VA
Total noncontinuous + continuous loads =	28,400 VA

*In the example, 125% of the actual connected lighting load (8500 VA × 1.25 = 10,625 VA) is less than 125% of the load from Table 220.12, so the minimum lighting load from Table 220.12 is used in the calculation. Had the actual lighting load been greater than the value calculated from Table 220.12, 125% of the actual connected lighting load would have been used.

Minimum Number of Branch Circuits Required

General Lighting: Branch-circuits need only be installed to supply the actual connected load *[see 210.11(B)]*.

$$8500 \text{ VA} \times 1.25 = 10,625 \text{ VA}$$

$$10,625 \text{ VA} \div 240 \text{ V} = 44 \text{ A for 3-wire, } 120/240 \text{ V}$$

The lighting load would be permitted to be served by 2-wire or 3-wire, 15- or 20-A circuits with combined capacity equal to 44 A or greater for 3-wire circuits or 88 A or greater for 2-wire circuits. The feeder capacity as well as the number of branch-circuit positions available for lighting circuits in the panelboard must reflect the full calculated load of 9000 VA × 1.25 = 11,250 VA.

Show Window

$$6000 \text{ VA} \times 1.25 = 7500 \text{ VA}$$

$$7500 \text{ VA} \div 240 \text{ V} = 31 \text{ A for 3-wire, } 120/240 \text{ V}$$

The show window lighting is permitted to be served by 2-wire or 3-wire circuits with a capacity equal to 31 A or greater for 3-wire circuits or 62 A or greater for 2-wire circuits.

Receptacles required by 210.62 are assumed to be included in the receptacle load above if these receptacles do not supply the show window lighting load.

Receptacles

Receptacle Load: 14,400 VA ÷ 240 V = 60 A for 3-wire, 120/240 V

The receptacle load would be permitted to be served by 2-wire or 3-wire circuits with a capacity equal to 60 A or greater for 3-wire circuits or 120 A or greater for 2-wire circuits.

Minimum Size Feeder (or Service) Overcurrent Protection *[see 215.3 or 230.90]*

Subtotal noncontinuous loads	12,200 VA
Subtotal continuous load at 125% (16,200 VA × 1.25)	20,250 VA
Total	32,450 VA

$$32,450 \text{ VA} \div 240 \text{ V} = 135 \text{ A}$$

The next higher standard size is 150 A *(see 240.6)*.

Minimum Size Feeders (or Service Conductors) Required *[see 215.2 and 230.42(A)]*

For 120/240-V, 3-wire system,

$$32,450 \text{ VA} \div 240 \text{ V} = 135 \text{ A}$$

Service or feeder conductor is 1/0 Cu per 215.3 and Table 310.16 (with 75°C terminations).

Figure 6–4 *(concluded)* (Reprinted with permission from NFPA 70-2005, the *National Electrical Code*®, Copyright © 2004–, National Fire Protection Association, Quincy, MA 02269. This reprinted material is not the complete and official position of the National Fire Protection Association on the referenced subject, which is represented only by the standard in its entirety.)

Example D8. Motor Circuit Conductors, Overload Protection, and Short-Circuit and Ground-Fault Protection *(see 240.6, 430.6, 430.22, 430.23, 430.24, 430.32, 430.52, and 430.62, Table 430.52 and Table 430.250)*

Determine the minimum required conductor ampacity, the motor overload protection, the branch-circuit short-circuit and ground-fault protection, and the feeder protection, for three induction-type motors on a 480-V, 3-phase feeder, as follows:

(a) One 25-hp, 460-V, 3-phase, squirrel-cage motor, nameplate full-load current 32 A, Design B, Service Factor 1.15

(b) Two 30-hp, 460-V, 3-phase, wound-rotor motors, nameplate primary full-load current 38 A, nameplate secondary full-load current 65 A, 40°C rise.

Conductor Ampacity

The full-load current value used to determine the minimum required conductor ampacity is obtained from Table 430.250 *[see 430.6(A)]* for the squirrel-cage motor and the primary of the wound-rotor motors. To obtain the minimum required conductor ampacity, the full-load current is multiplied by 1.25 *[see 430.22 and 430.23(A)]*.

For the 25-hp motor,

$$34 \text{ A} \times 1.25 = 42.5 \text{ A}$$

For the 30-hp motors,

$$40 \text{ A} \times 1.25 = 50 \text{ A}$$

$$65 \text{ A} \times 1.25 = 81.25 \text{ A}$$

Motor Overload Protection

Where protected by a separate overload device, the motors are required to have overload protection rated or set to trip at not more than 125% of the nameplate full-load current *[see 430.6(A) and 430.32(A)(1)]*.

For the 25-hp motor,

$$32 \text{ A} \times 1.25 = 40.0 \text{ A}$$

For the 30-hp motors,

$$38 \text{ A} \times 1.25 = 47.5 \text{ A}$$

Where the separate overload device is an overload relay (not a fuse or circuit-breaker), and the overload device selected at 125% is not sufficient to start the motor or carry the load, the trip setting is permitted to be increased in accordance with 430.32(C).

Branch-Circuit Short-Circuit and Ground-Fault Protection

The selection of the rating of the protective device depends on the type of protective device selected, in accordance with 430.52 and Table 430.52. The following is for the 25-hp motor.

(a) Nontime-Delay Fuse: The fuse rating is 300% × 34 A = 102 A. The next larger standard fuse is 110 A *[see 240.6 and 430.52(C)(1), Exception No. 1]*. If the motor will not start with a 110-A nontime-delay fuse, the fuse rating is permitted to be increased to 125 A because this rating does not exceed 400% *[see 430.52(C)(1), Exception No. 2(a)]*.

(b) Time-Delay Fuse: The fuse rating is 175% × 34 A = 59.5 A. The next larger standard fuse is 60 A *[see 240.6 and 430.52(C)(1), Exception No. 1]*. If the motor will not start with a 60-A time-delay fuse, the fuse rating is permitted to be increased to 70 A because this rating does not exceed 225% *[see 430.52(C)(1), Exception No. 2(b)]*.

Feeder Short-Circuit and Ground-Fault Protection

The rating of the feeder protective device is based on the sum of the largest branch-circuit protective device (example is 110 A) plus the sum of the full-load currents of the other motors, or 110 A + 40 A + 40 A = 190 A. The nearest standard fuse that does not exceed this value is 175 A *[see 240.6 and 430.62(A)]*.

Figure 6–5 Examples of motor calculations. (Reprinted with permission from NFPA 70-2005, the *National Electrical Code®*, Copyright © 2004–, National Fire Protection Association, Quincy, MA 02269. This reprinted material is not the complete and official position of the National Fire Protection Association on the referenced subject, which is represented only by the standard in its entirety.)

specific requirements throughout the latter chapters in the *NEC®*. Several examples for calculating branch-circuits are found in the latter sections of this chapter. Article 215 covers the installation requirements and minimum size and ampacity for conductors for feeders supplying branch-circuit loads, as calculated in Article 220.

Example: Is it permissible for the required single 20-ampere laundry circuit in a dwelling to supply both an automatic washer with a nameplate of 9.8 amperes and a gas dryer with a nameplate of 7.2 amperes?

Step 1. *NEC®* 210.52(F) requires at least one receptacle outlet.

Step 2. *NEC®* 210.11(C)(2) requires at least one 20-ampere circuit for laundry outlet(s) to supply the receptacle(s) required by 210.52(F); 422.10(B); also see references in 210.23.

> *Note:* This means all outlets for laundry purposes, that is, washer, 120-volt dryer, and ironing closet.

Step 3. *NEC®* 210.24 and Table 210.24 reference 210.23(A).

Step 4. *NEC®* 210.23(A); the rating of any one utilization equipment supplied from a cord and plug shall not exceed 80 percent of the branch-circuit rating.

Step 5. 20 × 80 percent = 16 amperes. Therefore, that requirement is acceptable.

Step 6. 9.8 + 7.2 = 17 amperes. The rating of the circuit is not exceeded.

Answer: Yes. Reference: 422.10(B), 210.24, 210.23, 210.23(A)(1), and 210.11(C)(2). Also, see definition of "**continuous load**." These would not be considered a continuous load.

Outside Feeders and Branch-Circuits

Article 225 covers outside branch-circuits and feeders. Article 225 covers the electrical equipment and wire for the supply of utilization equipment located on or attached to the outside of public and private buildings, or run between buildings, other structures, or poles on premises served. For clearances, see Figure 6–6. An example of a feeder would be those conductors being fed from an overcurrent device in the service equipment to a subpanel in another part of the building or feeding a separate building on the premises, terminating in a panel or group of overcurrent devices within the second building or in another location. A feeder also serves separately derived systems in many cases, such as a transformer where the voltage is reduced to serve lighting and receptacle branch-circuit loads.

NEC® 225.32 provides that where more than one building or structure is on the same property and under single management, each building or other structure served shall be provided with a means of disconnecting all ungrounded conductors, which must be installed either inside or outside the building or structure in a readily accessible location nearest the point of entrance of the supply feeder conductors. The disconnect disconnecting these feeder conductors must be installed in accordance with 230.70 and 230.72 and must be suitable as service equipment. However, it is required to have overcurrent protection in accordance with Article 220 for branch-circuits and Article 215 for feeders. The reference to Article 220 is necessary because residential buildings and structures such as garages and other outbuildings are permitted to be supplied with branch-circuits where the loads permit. Examples for calculating feeders and outside branch-circuits are shown in this chapter.

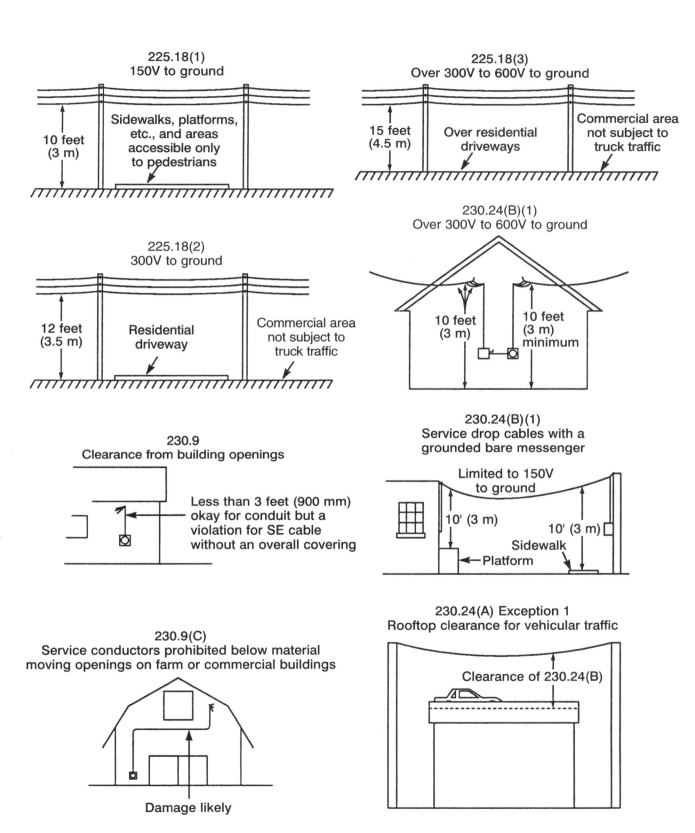

Figure 6–6 Required clearances for service and feeder conductors.

QUESTION REVIEW

Each lesson is designed purposely to require the student to apply the entire *NEC*® text, and not specific chapters or articles. It has been found that when studying for a timed, open-book examination, the student must gain proficiency in the Table of Contents, the Index, and the ability to move quickly from cover to cover to find the correct answer to each question in a timely fashion.

1. Fluorescent luminaires (lighting fixtures), each containing two ballasts rated 0.8 ampere each at 120 volts, are to be installed for general lighting in a store. The overcurrent protection devices are not listed for continuous operation at 100 percent of its rating. What is the maximum number of these luminaires (lighting fixtures) that may be permanently wired to a 20-ampere, 120-volt branch-circuit?

Answer: _____

_____

_____

_____

Reference: _____

The following drawings apply to Questions 2 through 20.

Example for Questions 2 through 20: You are going to build an investment single-family dwelling 28 feet by 40 feet (8.5 m by 12 m). There will be three bedrooms and one bathroom with central natural gas heat located under the floor in the crawl space. There will be no air conditioning. There will be an electric water heater (40-gallon, quick recovery) with two 4500-watt elements, a gas cooking range, and electric dryer. There will be a detached single-car garage. The electrical installation will be made in accordance with the 2005 *NEC*®. The attic in this dwelling is not suitable for the location of equipment or for storage.

2. How many square feet (square meters) will this house contain?

   Answer: _____

   _____

   _____

   _____

   Reference: _____

3. How many 15-ampere general lighting and receptacle branch-circuits are required?

   Answer: _____

   _____

   _____

   _____

   Reference: _____

4. How many 20-ampere lighting and receptacle branch-circuits are required?

   Answer: _____

   _____

   _____

   _____

   Reference: _____

5. How many appliance branch-circuits are required?

   Answer: _____

   _____

   _____

   _____

   Reference: _____

6. Is a special circuit required for the laundry?

   Answer: _____

   _____

   _____

   _____

   Reference: _____

7. What is the maximum standard size overcurrent device permitted to protect the electric water heater?

   Answer: _____

   _____

   _____

   _____

   Reference: _____

8. How many branch-circuits are required to supply the detached garage?

   Answer: _____

   _____

   _____

   _____

   Reference: _____

9. How many outdoor receptacles are required?

   Answer: _____

   _____

   _____

   _____

   Reference: _____

10. What is the minimum number of ground-fault circuit interrupter (GFCI) devices required?

   Answer: _____

   _____

   _____

   _____

   Reference: _____

11. Are lighting outlets required at each outside door?

Answer: _____

_____

_____

_____

Reference: _____

12. Are lighting outlets required in the attic not used for storage?

Answer: _____

_____

_____

_____

Reference: _____

13. Is a receptacle required under the floor containing equipment requiring servicing?

Answer: _____

_____

_____

_____

Reference: _____

14. If the answer to Question 6 is yes, what size branch-circuit would be required?

Answer: _____

_____

_____

_____

Reference: _____

15. Would it be permissible to connect the furnace to a general lighting and receptacle branch-circuit provided it does not overload the circuit?

Answer: _____

_____

_____

_____

Reference: _____

16. Are lighting outlets required in each closet?

Answer: _____

_____

_____

_____

Reference: _____

17. Is it permissible to connect the electric ignition system for gas ranges to one of the required small-appliance branch-circuits?

Answer: _____

_____

_____

_____

Reference: _____

18. If an outdoor receptacle is located 6 feet (1.8 m) above the ground level, is it required to be GFCI protected?

Answer: _____

_____

_____

_____

Reference: _____

19. Could the outdoor receptacle described in Question 18 serve as one of the required outdoor receptacle?

Answer: _____

_____

_____

_____

Reference: _____

20. What is the minimum size service required for this dwelling? (Calculate answers where required.)
    (a) General purpose lighting and receptacle load
    (b) Appliance branch-circuit load
    (c) Water heater
    (d) Laundry
    (e) Dryer
    (f) Detached garage
    (g) Furnace
    (h) Outdoor receptacle
    (i) Air conditioning
    (j) Range
    (k) Bathroom

(a) _____

(b) _____

(c) _____

(d) _____

(e) _____

(f) _____

(g) _____

(h) _____

(i) _____

(j) _____

(k) _____

Chapter Seven

OBJECTIVES

Studying this chapter along with the 2005 *NEC®* gives the reader a basic introduction to the service requirements in the *Code* that apply to all installations for above and below 600 volts.

After studying this chapter, you should know:

- The definitions and the application of terms that are applicable and are unique to the 2005 *NEC®*
- How to find the many *Code* requirements in which services are addressed as they apply to the *NEC®*
- Service circuit sizing
- Service conductor sizing
- Service overcurrent sizing
- Service raceway sizing
- How to find the *Code* requirements in which high-voltage services are addressed as they apply to the *NEC®*

SERVICES 600 VOLTS OR LESS

NEC® Article 230 covers the requirements for services, service conductors, and equipment for the control and protection of services and their installation requirements. Parts I through VII cover services, 600 volts, nominal, or less; Part VIII has additional requirements for services exceeding 600 volts, nominal. It should be remembered that all of Article 230 applies to these services, and Part VIII is additional provisions that supplement or modify the rest of Article 230. Part VIII shall not apply to the equipment on the supply-side of the service point. Clearances for conductors over 600 volts are found in ANSI C2, the *National Electrical Safety Code. NEC®* 230.2 generally limits a building or structure to one service. There are several special conditions that permit more than one service per building or structure. It is necessary to comply with one of these special conditions any time two or more services are being installed on a building or structure.

Confusing to many is the permissiveness for more than one set of service conductors permitted by the Exceptions of 230.40. These exceptions frequently are used on strip shopping centers, condominium complexes, and apartment buildings. The rule satisfies the needs for individual supply, control, and metering of occupancies in these multi-occupancy buildings. However, it does not relieve the requirement of 230.2 for one service for each building or structure (see Figure 7–1 and Figure 7–2). The service conductors, as defined in Article 100, are required to go directly to the disconnecting means of that building or structure, because the disconnecting means is required to be installed at a readily accessible location either outside of the building or structure, or inside nearest the point of the entrance of the service conductors. *NEC®* 230.6 defines that conductors shall be considered outside of a building or other structure where installed under not less than 2 inches (50 mm) of concrete beneath a building or structure, where installed within a building or structure in a raceway that is encased in concrete or brick not less than 2 inches (50 mm) thick, or where installed in a vault that complies with the requirements of *NEC®* Article 450, Part III. A typical service supplying a building or structure contains several carefully defined parts. These parts are defined in Article 100 as service, service cable, service conductors, service drop, service-entrance conductors, overhead system, underground system, service lateral, service equipment, and service point. Each service will contain several of these parts—however, not all of these parts. It is important that the components being installed in accordance with the *NEC®* be correctly defined before sizing, calculating, or installing. For instance, a typical residential service can be supplied by the serving utility either overhead or underground. When supplied overhead, it shall be supplied as a service drop. When supplied underground, it shall be supplied as a service lateral. These terms should be studied carefully in *NEC®* Article 100 so that calculations and design can be made correctly. The diagram in figure 230.1 should help the reader identify and use these terms correctly. (See Figure 7–3.)

Electrical service components

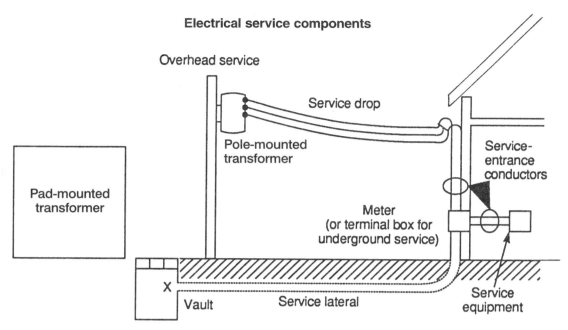

Figure 7–1 *NEC*® 230.2 permits only one service to each building. There are several conditions to this rule where more than one service is permitted. The service may be supplied with aerial spans overhead or by underground "service laterals" from the "service point." (See Article 100—Definitions.) *Note:* A vault is not required for underground runs but may be installed where necessary.

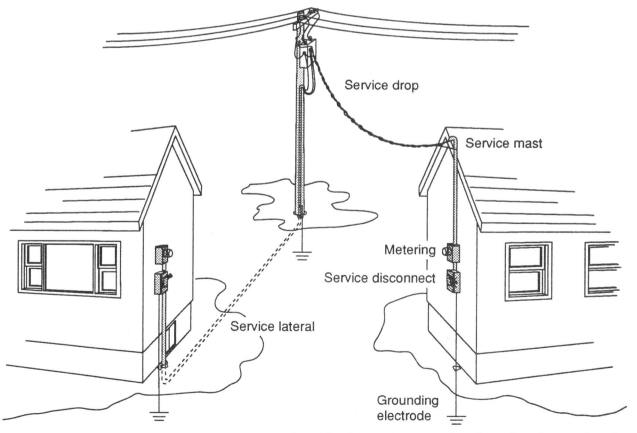

Figure 7–2 Illustration of two common types of residential services, overhead service drop and underground service lateral.

Article 230 – Services

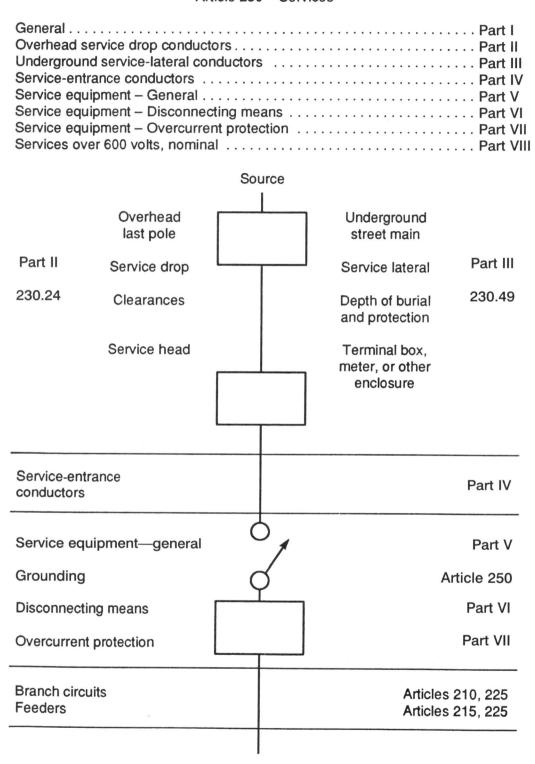

Figure 7–3 Article 230 is divided into several parts. Each covers a specific part of service installation. (Reprinted with permission from NFPA 70-2005, the *National Electrical Code®*, Copyright © 2004–, National Fire Protection Association, Quincy, MA 02269. This reprinted material is not the complete and official position of the National Fire Protection Association on the referenced subject, which is represented only by the standard in its entirety.)

Part II of Article 230 covers the overhead service-drop conductors, size and rating, and clearances required for making this installation. These conductors are generally installed by the serving utility; however, it is the responsibility of the installer to locate the service mast so that these clearances will be met. Careful and close coordination between the serving utility, the customer to be served, and the installer is necessary to ensure compliance with this section of the *NEC*®. (Refer to Figure 6–6 for clearance examples.)

Part III of Article 230 covers underground service-laterals. (See Figure 7–4.) Care should be taken so that the service-lateral is installed correctly and so that where it emerges from the ground it is amply protected by rigid or IMC steel conduit or Schedule 80 PVC to provide these conductors with physical protection against damage. Article 230, Part IV covers the requirements for the service-entrance conductors. These service-entrance conductors, as defined in *NEC*® Article 100, must be installed in one of the wiring methods listed in 230.43, and sized and rated in accordance with 230.42. Specific requirements for these service-entrance conductors are covered in Article 230, Part IV and must be complied with regardless of how they are routed. The general requirements for service equipment can be found in Article 230, Part V, and the disconnecting means is found in Part VI.

Each service disconnecting means is required to be suitable for the prevailing equipment conditions, and when installed in hazardous locations must comply with Chapter 5 of the *National Electrical Code*®. Each service disconnecting means permitted by 230.2 or for each set of service-entrance conductors permitted by 230.40, Exception 1 (see Figure 7–5) shall consist of not more than six switches or six circuit breakers mounted in a single enclosure, in a group of separate enclosures, or on a switchboard. (See Figure 7–6.) There shall be no more than six disconnects per service grouped in any one location. These requirements are critical and must be followed. The disconnecting means must be grouped; each disconnect must be marked to indicate the load served. An exception permits one of the six disconnecting means permitted where used for a water pump intended to provide fire protection to be located remotely from the other disconnecting means. In multiple-occupancy buildings, each occupant shall have access unless local building management, which supplies continuous supervision, is present. In such a case, the service disconnecting means shall be permitted to be accessible to the authorized management personnel only. Each service disconnecting means shall simultaneously disconnect the ungrounded service conductors from the premise wiring. Where the service disconnecting means does not disconnect the grounded conductor

Electrical service components

Overhead service

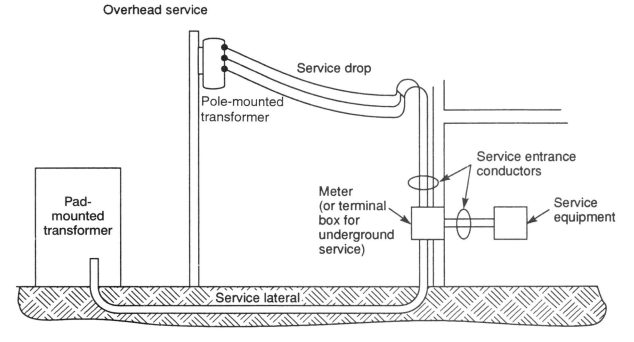

Figure 7–4 *NEC*® 230.2 permits only one service per building. It is permissible to supply overhead or underground.

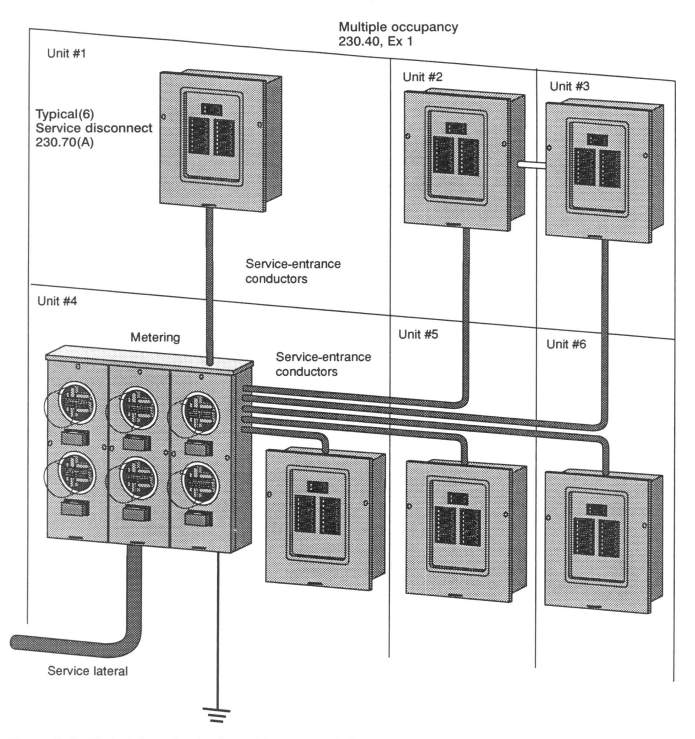

Figure 7–5 Typical riser sketch of a multioccupancy building. *NEC*® 230.40 Exception 1 permits the main disconnecting means for that occupancy to be located on the load end of the feeder within the occupancy.

from the premise wiring, other means shall be provided for this purpose in the service equipment. A terminal or bus to which all grounded conductors can be attached by means of pressure connections shall be permitted for this purpose.

Example: A 277/480-volt, 3-phase service to a building is rated at 1600 amperes. The required main disconnecting means consists of two 800-ampere fused disconnects. Is this service required to have ground-fault protection (GFP)?

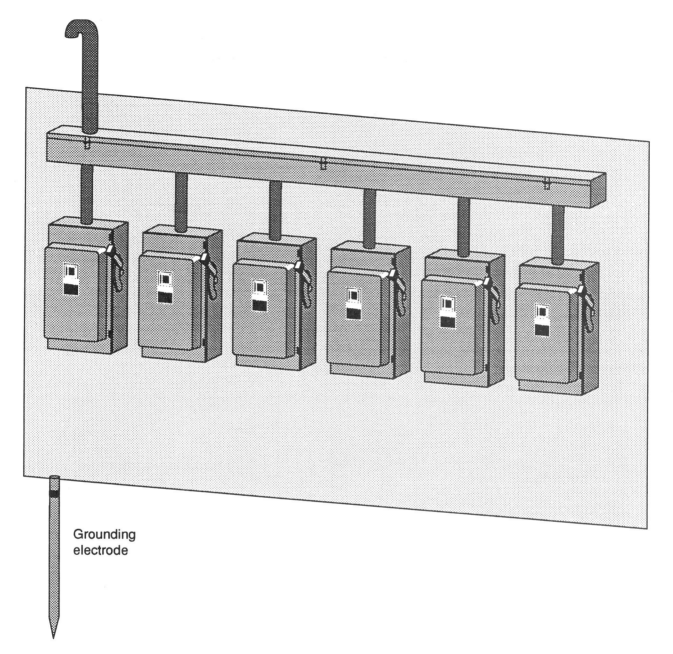

Figure 7–6 Typical sketch of six individual disconnecting means on one service as permitted by *NEC®* 230.71(A).

Index "Ground-Fault Protection," *Ground-fault protection for service disconnects, 230.95.*

Note: To answer this question, some basic electrical knowledge is required. First, the voltage tells the story. The number 277/480 indicates it is first a solidly grounded system. It also indicates it is a wye system because 480 equals 173% of 277. This information can be obtained from additional reference books listed as study material.

Answer: No. References: 230.71, the maximum number of service disconnects six (6). *NEC®* 230.95 states that solidly grounded wye services of more than 150 volts to ground not exceeding 600 volts phase-to-phase and over 1000 amperes requires ground-fault protection.

Part VII covers the overcurrent protection and the location of the overcurrent protection requirements for services. All solidly grounded wye services of more than 150 volts to ground but not exceeding 600 volts phase-to-phase rated 1000 amperes or more must be provided with ground-fault protection of equipment. Two exceptions to this: Exception 1 states that ground-fault protection does not apply to the service disconnecting means of a continuous industrial process where a nonorderly shutdown will introduce additional or increased hazard. Exception 2 states that it does not apply to fire pumps. Specific settings and testing procedures for this ground-fault protection of equipment are covered in 230.95. Additional ground-fault protection of feeders where ground-fault protection has not been provided on the service is covered in both Article 215 and Article 240.

Example: Where a steel conduit contains eight (8) 3/0 AWG THWN insulated copper conductors with two of the 3/0 AWG conductors being in parallel for each phase, and two 3/0 AWG conductors being in parallel for the neutral, (a) would this installation be permitted to be terminated into a 400-ampere main service breaker where the system voltage is 120/208? (The expected loads are evenly divided between fluorescent lighting, data processing equipment, and air-conditioning equipment.) (b) If not, what would be the maximum ampere rating for this service?

Answer: Eight 3/0 AWG THWN conductors in a single conduit, Annex C, Table C8, would require a trade size 3 (78) rigid metal conduit.

Table 310.16—3/0 AWG THWN has an ampacity of 200 amperes (from 75°C column). Where the number of current-carrying conductors exceeds three, the allowable ampacity shall be reduced as shown in the table with seven to nine conductors, a derating of 70 percent shall be applied. See Table 310.15(B)(2).

Two conductors per phase:

$$2 \times 200 \times 70\% = 280 \text{ amperes}$$

The next higher overcurrent device per 240.4 would be 300 amperes. The service would be a 300-ampere service.

The 400-ampere overcurrent device would require 2-350 kcmil per phase and with eight 350 kcmil THWN would require the rigid metal conduit to be increased to trade size 4 (103). For conductor termination, see 110.14(C).

For services exceeding 600 volts, nominal, Part VIII must be read and adhered to carefully. The definition of service point formerly appeared in this section. The examples given in this section help clarify the application of *NEC*® Article 230.

Example: A trade size 3 (78) steel intermediate metal conduit (IMC) contains 120/208 3-phase service conductors consisting of eight size 3/0 AWG THWN copper conductors. Each phase consists of two conductors per phase, and the full-size, neutral (grounded conductor) consists of two conductors per phase terminating in a 400-ampere service-disconnecting means consisting of an inverse time circuit breaker. The data processing, fluorescent lighting, and air-conditioning load is evenly divided. Does this installation meet *Code*?

Step 1. Chapter 9 Table 1 permits a 40 percent fill of this conduit. For 3/0 AWG type THWN conductors, Annex C Table C4A would allow eight 3/0 AWG THWN conductors in a trade size 2½ (63) conduit. Therefore, the trade size 3 (78) IMC conduit is acceptable.

> *Note:* Due to the change in the tables in Chapter 9 and the addition of Annex C in the *NEC*®, each type of raceway must be calculated individually. For example, in this problem if nonmetallic conduit Schedule 40 were used, a trade size 3 (78) would be required. If more than one size conductor was contained within the raceway, then calculations would be necessary using Chapter 9, Tables 1, 4, and 5 for the number of conductors, the specific type and size, and the specific raceway type.

Step 2. The ampacity of the 3/0 AWG THWN conductors is 200 amperes each based on Table 310.16. However, Table 310.15 (B)(2) requires a deration to 70 percent where eight conductors are in one raceway. So, 200 amperes × 2 = 400 × 70% = 280-ampere ampacity. 230.90(A) Exception 2 references 240.4, (C), and 240.6, which permits overload protection of the conductors at the next higher standard size overcurrent device in accordance with 240.6. Thus, the largest size overcurrent device permitted would be a 300-ampere device.

Step 3. Assuming the calculated load required the 400-ampere device, you would be required to either increase the conductor size, which would indubitably require a larger conduit or two conduits in parallel with conductors sized to carry the calculated load after derating.

Step 4. A 400-ampere calculated load would require a trade size 3 (78) IMC and a trade size 4 (103) PVC Schedule 40, nonmetallic raceway to contain the eight required 350-kcmil type THWN conductors.

Answer: Table 310.16 and Table 310.15(B)(2): $310 \times 2 = 620 \times 70\% = 434$.

SERVICES OVER 600 VOLTS

Services over 600 volts nominal introduce an additional level of hazard. Article 230, Part VIII covers the requirements for services exceeding 600 volts nominal, which modify or amend the rest of Article 230. All of Article 230 is applicable and, in addition, Part VIII must be followed. Clearance requirements for over-600-volt services can be governed by ANSI C2, the *National Electrical Safety Code*, as noted in the FPN in 230.200. The service-entrance conductors to buildings or enclosures shall not be smaller than 6 AWG unless in cable, and in cable not smaller than 8 AWG, and installed by the specific methods listed in 230.202(B). It should be noted that Article 300, Part II and Article 110, Part II also will need to be regularly referenced in making installations over 600 volts. The requirements for support guarding and draining cables are covered in this section. Warning signs with the words "Danger. High Voltage. Keep Out." shall be posted where unauthorized people come in contact with energized parts. The disconnecting means must comply with 230.70 or 230.208(B) and shall simultaneously disconnect all underground conductors and shall have fault closing rating of not less than the maximum short circuit available in supply terminals, except where used switches or separate mining fuses are installed, the fuse characteristic shall be permitted to contribute to the fault closing rating of the disconnecting means. These requirements for over-600-volt services create the need for your studies to take you into new areas of the *Code*, such as Part II of Article 100, which covers the general requirements for over 600 volts nominal; Part II of Article 100, which gives you the unique definitions for those circuits; Part IX of Article 240; Part XI of Article 250 for grounding of system; and Part II of Article 300; and the requirements of Article 490. You must be aware that these additional requirements amend or augment the primary requirements in each article. In most instances, those requirements also apply to over-600-volt nominal services. Therefore, as you study this section, familiarize yourself with these new parts as they apply to your studies.

QUESTION REVIEW

Each lesson is designed purposely to require the student to apply the entire *NEC®* text, and not specific chapters or articles. It has been found that when studying for a timed, open-book examination, the student must gain proficiency in the Table of Contents, the Index, and the ability to move quickly from cover to cover to find the correct answer to each question in a timely fashion.

1. Where a circuit breaker is provided as the short-circuit protective device for service-entrance conductors exceeding 600 volts, the breaker shall have a trip setting of not more than _____ times the ampacity of the conductors.

Answer: _____

_____

_____

_____

Reference: _____

2. Where a group of buildings are served by electrical feeders from a single service drop or electrical service point in another building, is a disconnecting means required at each of the other buildings where the feeders terminate or can the feeders terminate in main lug-only panels?

Answer: _____

_____

_____

_____

Reference: _____

3. Is there a limit to the number of overcurrent devices that can be used on the secondary side of a transformer if the total of the device ratings does not exceed the allowed value for a single secondary overcurrent device?

Answer: _____

_____

_____

_____

Reference: _____

4. Does it make a difference which opening in the service head the service-entrance conductors are brought out of? If so, why?

Answer: _____

_____

_____

_____

Reference: _____

5. If the clearances required by *NEC®* 110.26 can be maintained in front of a panelboard or a service panel to be installed in a bathroom in a dwelling, would this installation be permitted by the *NEC®*?

Answer: _____

_____

_____

_____

Reference: _____

6. When service-entrance cable is used within a dwelling for branch-circuit wiring, is the Type SE cable considered the same as Type NM nonmetallic-sheathed cable?

Answer: _____

_____

_____

_____

Reference: _____

7. Can a 120/240-volt, single-phase, 3-wire service be located closer than 3 feet (900 mm) to a window of a dwelling?

Answer: _____

_____

_____

_____

Reference: _____

8. The *Code* generally prohibits splicing a grounding-electrode conductor. (a) Is this conductor permitted to be tapped? (b) If so, how are these taps sized?

Answer: _____

_____

_____

_____

Reference: _____

9. Circuit breakers are required to open all ungrounded conductors of a circuit simultaneously. Obviously, fuses cannot be required to open all ungrounded conductors of a circuit simultaneously. Why the difference in the requirements?

Answer: _____

_____

_____

_____

Reference: _____

10. What is a separately derived system? Where is it mentioned in the *Code*?

Answer: _____

_____

_____

_____

Reference: _____

11. How many overcurrent devices are permitted to be used on the secondary side of a transformer if the total of the device ratings does not exceed the allowed value for a single overcurrent device?

Answer: _____

_____

_____

_____

Reference: _____

12. Is it necessary to identify the higher voltage-to-ground phase (hi-leg) at the disconnect of a 3-phase motor where the service available is 120/240-volt, 3-phase, 4-wire?

Answer: _____

_____

_____

_____

Reference: _____

13. A service panel is securely bolted to the metal frame of a building that is effectively grounded, and the panelboard's main bonding jumper is properly installed. Is this sufficient to ground the panelboard to the metal building grounding electrode as required in *NEC*® 250.64?

Answer: _____

_____

_____

_____

Reference: _____

14. What is the maximum permitted contraction or expansion for a run of Schedule 40 PVC conduit before the installation of an expansion coupling is required?

Answer: _____

_____

_____

_____

Reference: _____

15. On a recent installation, the plans called for the building steel to be used as a grounding electrode if effectively grounded. What does this term mean, and where is it used in the *NEC*®?

Answer: _____

_____

_____

_____

Reference: _____

16. Which article of the *NEC*® covers Type nonmetallic-sheathed cable?

Answer: _____

_____

_____

_____

Reference: _____

17. What is the maximum size flexible metallic tubing available manufactured today?

    Answer: _____

    _____

    _____

    _____

    Reference: _____

18. Which article of the *National Electrical Code*® contains the requirements for fusible safety switches and other type snap switches?

    Answer: _____

    _____

    _____

    _____

    Reference: _____

19. In making an installation in a residential garage, you have determined that physical protection is needed for the nonmetallic-sheathed cable and that it must be installed in steel conduit. How do you size the conduit for this nonmetallic-sheathed cable?

    Answer: _____

    _____

    _____

    _____

    Reference: _____

20. In making an installation for a large parking garage adjacent to a hotel building, which article of the *NEC*® covers these wiring methods?

    Answer: _____

    _____

    _____

    _____

    Reference: _____

Chapter Eight

OBJECTIVES

Studying this chapter along with the 2005 *NEC®* gives the reader a basic introduction to the overcurrent requirements in the *Code* that apply to all installations for above and below 600 volts.

After studying this chapter, you should know:

- The definitions and the application of terms that are applicable and are unique to the 2005 *NEC®*
- The many *Code* articles in which overcurrent is addressed as it applies to the *NEC®*
- Overcurrent sizing
- Fuse applications and types
- Circuit breaker types and applications
- Short circuit/fault current calculations

OVERCURRENT PROTECTION

NEC® Article 240 provides the general requirements for overcurrent protection and overcurrent protective devices not more than 600 volts, nominal. Part IX of *NEC®* Article 240 covers the overcurrent protection for over 600 volts, nominal. The overcurrent protection for conductors and equipment is provided to open the circuit if the current reaches a value that will cause an excessive or dangerous temperature in conductors or conductor insulation. *NEC®* 110.9 and 110.10 cover the requirements for the interrupting capacity and protection against fault currents (see Figure 8–1). Article 240 covers the general requirements. The specific overcurrent requirements for equipment can be found in each individual article as it pertains to that specific equipment or circuitry; for example, the overcurrent protection for air-conditioning and refrigeration equipment is covered in Article 440 and appliances in Article 422. A listing of the specific overcurrent requirements can be found in 240.3. *NEC®* 240.4 outlines the protection of conductors. This section provides the rules for the many different applications throughout the *Code. NEC®* 240.5 covers the requirements for protecting flexible cords and fixture wires,

Figure 8–1 Panelboards. *(Courtesy of Square D Company)*

and standard overcurrent devices are listed in *NEC®* 240.6 for fuses and fixed-trip circuit breakers. See Figure 8–2 and Figure 8–3 for examples. Adjustable-trip circuit breakers are described and covered in 240.6(B). Part II of Article 240 specifies the location that the overcurrent device must be placed in the circuit. Part III specifies the enclosure required for enclosing the overcurrent devices. Part IV covers the disconnecting requirements for disconnecting the overcurrent devices so that they are accessible to the maintenance personnel performing work on those devices. Parts V and VI cover fuses, fuseholders, and adapters for both plug- and cartridge-type fuses; circuit breakers are covered in Part VII. Article 240, Part VIII covers supervised industrial installations. Part IX

Figure 8–2 Types IFL, IKL, and ILL "1 LIMITER" molded case current-limiting circuit breakers. *(Courtesy of Square D Company)*

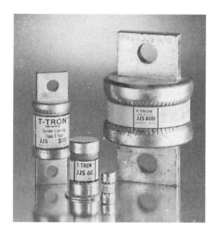

Figure 8–3 Class T current-limiting, fast-acting fuse; 200,000-ampere interrupting rating, 300 and 600 volts. Has little time delay. Generally used for protection of circuit-breaker panels and for circuits that do not have high-inrush loads, such as motors. When used on high-inrush loads, generally size at 300 percent so as to be able to override the momentary inrush current. These fuses have different dimensions compared with ordinary fuses. They will not fit into switches made for other classes of fuses, nor will other classes of fuses fit into a Class T disconnect switch. *(Courtesy of Cooper Bussmann)*

covers the overcurrent protection for over 600 volts, nominal, and states that feeders shall have short-circuit protection devices in each ungrounded conductor or comply with 230.208(A) and (B). The protective device(s) shall be capable of detecting and interrupting all values of current that can occur at their location in excess of their trip setting or melting point. In no case shall the fuse rating exceed three times the long-time trip element setting of the breaker, or six times the ampacity of the conductor. Branch-circuit requirements for over-600-volt circuits are covered in 240.100.

Example: What is the maximum size time-delay fuse permitted to protect a 1-horsepower, 3-phase, 230-volt squirrel cage motor with a nameplate of 3.3 amperes?

Step 1. *NEC®* 430.6. Use the tables and not the nameplate amperes.

Step 2. Table 430.150. One horsepower is 3.6 amperes.

Step 3. *NEC®* 430.52(A), Exception 1. (Where the standard overcurrent device size does not correspond as permitted by Table 430.152, the next higher is permitted.)

Step 4. Table 430.152 permits 175 percent for a time-delay fuse. 3.6 amperes × 175% = 6.3 amperes

Step 5. *NEC®* 240.6. The next higher standard fuse is 10 amperes. Therefore, a 10-ampere time-delay fuse would be permitted.

Answer: 10 amperes

Example: A 120/208V, 3-phase lighting and branch-circuit panelboard is to be fed with 23-foot long copper tap conductors tapped from three 350-kcmil THWN copper conductors run in parallel protected by an 800-ampere circuit breaker. The panelboard supplies a continuous-duty luminaire (lighting) load of 135 amperes per phase; one hundred five 120-volt receptacle outlets; four 3-horsepower, 3-phase squirrel-cage motors; and one 25-horsepower, 3-phase, 5-minute-rated, short-time-duty squirrel-cage motor. What are the minimum size tap conductors required?

Step 1. Index "Taps." See *Feeders, Feeder taps,* 240.21 and 430.28.

Step 2. Although motors are a portion of the intended load, these taps are to feed a panelboard with mixed loads.

Step 3. See 240.21(C). Because 23 feet is more than 10 and less than 25, we must use the 25-foot rule.

Step 4. The ampacity of the tap conductors must be a minimum of one third the rating of the overcurrent device protecting the circuit. Therefore, 800 × ⅓ = 267 amperes. Reference Table 310.16. THWN Copper 75°C would require a minimum tap conductor size of 300 kcmil. Now, calculations are necessary to see whether the load exceeds the minimum. If it does, then the larger of the two would be required.

Continuous duty lighting 135 × 125%   = 168.75
(220.42)

105 receptacles @ 180 VA = 18,900 VA
@ 10,000 + (8900 × 50%) =
14,450 ÷ 120   = 120.4
(220.14(I) Table 220.42)

Three 3 hp motors @ 10.6   =   1.8
(430.6, Table 430.150)

One 25 hp motor 5-minute short-time
duty. 74.8 × 110%   = 82.28
(430.24 Exception 1,
430.22 Exception 1, and
Table 430.150)

Round to the nearest whole number   = 403

Answer: The minimum size is not adequate. Therefore, 600 kcmil per phase with a 420 ampacity would be required (310.16).

Example: What is the minimum size copper THW feeder required for a single-phase 120/240-volt office building feeder with the maximum unbalanced load of 125 amperes of continuous-duty fluorescent lighting on each ungrounded conductor?

Step 1. Index "Feeder Calculation of Loads." *NEC®* 220.40 and Chapter 9 Part II.
Step 2. 220.11 requires that continuous-duty loads be calculated at the load plus 25 percent for sizing the overcurrent device.
Step 3. 125 × 125% = 146.25 amperes. See 240.4 (B) = 150-ampere overcurrent device.
Step 4. 310.15, Table 310.16 Ampacity tables.

Answer: 1/0 AWG copper-type THW conductors

Example: Provide the fuse sizes for a 10-horsepower, 230-volt, 3-phase motor code lettered G, full voltage start with a service factor of 1.15.

(A) The fuse sizes using time-delay fuses for the branch-circuit protection is calculated as follows: 28 × 1.75 = 49 amperes. (*NEC®* 430.52 requires that you round down to the next lower standard size. See 240.6.) Therefore, install 45-ampere time-delay fuses. If the 45-ampere time-delay fuses are not sufficient to carry the load or to start the motor, a larger size time-delay fuse may be installed, but not to exceed 225 percent.

(B) Using time-delay fuses for motor overload protection, the size will be as follows: 28 × 1.25 = 35-ampere time-delay fuses.

(C) Using nontime-delay fuses, sizes for branch-circuit and ground-fault protection are calculated as follows: 28 × 3 = 84 amperes. Round down to 80-ampere nontime-delay fuses as explained in (A). This would permit a maximum size of 28 × 4 = 112. Install 110-ampere nontime-delay fuses. References for making these calculations are Table 430.150 and Table 430.52, 240.6, 430.32, and 430.52.

Example: To calculate the available short-circuit current at a panelboard located 20 feet (6 m) away from a transformer when the transformers are 500 kcmil copper in steel conduit. The "C" value for the conductors is 22,185. The transformer is marked 300 kVA, 208/120 volts, 3-phase, 4-wire. The transformer impedance is 2 percent. Consider the source to have an infinite amount of fault current available (often referred to as "infinite primary").

(1) Find the transformer's full-load secondary current:

$$I_{fla} = \frac{kVA \times 1000}{E \times 1.73} = \frac{300 \times 1000}{208 \times 1.73} = 834 \text{ amperes}$$

(2) To find the transformer SCA:
(multiplier = 100/2 = 50)

Transformer SCA =

$I_{fla} \times$ multiplier = 834 × 50 = 41,700 amperes

(3) To find the (F) factor:

$$F = \frac{1.73 \times L \times I}{C \times E_{L-L}} = \frac{1.73 \times 20 \times 41,700}{22,185 \times 208} = 0.3127$$

(4) To find the (M) multiplier:

$$M = \frac{1}{1+F} = \frac{1}{1+0.3127} = 0.76$$

(5) To find the short-circuit current of the panelboard:

$$I_{sca} = transformer_{sca} \times (M) = 41,700 \times 0.76 = 31,692 \text{ amperes}$$

The fault current available at the panelboard where it is located 20 feet (6 m) away from the transformer is equal to 31,692 amperes.

QUESTION REVIEW

Each lesson is designed purposely to require the student to apply the entire *NEC®* text, and not specific chapters or articles. It has been found that when studying for a timed, open-book examination, the student must gain proficiency in the Table of Contents, the Index, and the ability to move quickly from cover to cover to find the correct answer to each question in a timely fashion.

1. Does the *Code* permit the use of two single-pole circuit breakers in a panelboard to serve a line-to-line connected load such as a household electrical range rated at 120/240 volts or a hot water heater rated at 240 volts?

Answer: _____

_____

_____

_____

Reference: _____

2. Where outdoor conductors are tapped and these tapped conductors terminate into a single overcurrent device designed to limit the load so as not to exceed the ampacity of the tap conductors, how long can these tap conductors be at the maximum length?

Answer: _____

_____

_____

_____

Reference: _____

3. Is it permissible to use 300-volt cartridge-type fuses and fuseholders to protect a multiwire, 4-wire circuit such as an emergency circuit in a food store where the electrical service is 277/480-volt, 3-phase, 4-wire?

Answer: _____

_____

_____

_____

Reference: _____

4. Service-entrance conductors are required to be protected at their rated ampacity with exception, the exception being the next standard size overcurrent device when the service conductors are protected by a single device. When are multiple overcurrent devices supplied by a service allowed to exceed the ampacity rating of the service-entrance conductors?

Answer: _____

_____

_____

_____

Reference: _____

5. Fixed electric space heating loads shall be computed at _____ percent of the total connected load, generally.

Answer: _____

_____

_____

_____

Reference: _____

6. A motor is to be installed in a Class I, Division 2 location. It is to be connected with 3 feet (900 mm) of trade size ¾ (21) liquidtight flexible metal conduit. Is it permissible to use the liquidtight flexible metal conduit as the grounding path for this installation?

Answer: _____

_____

_____

_____

Reference: _____

7. Are feeder conductors required to be rated for 125 percent of the continuous load plus the noncontinuous load?

Answer: _____

_____

_____

_____

Reference: _____

8. How many duplex receptacles can be installed on a 20-ampere branch-circuit in a dwelling?

   Answer: _____

   _____

   _____

   _____

   Reference: _____

9. Table 310.16 lists the allowable ampacity of 12 AWG THWN conductors as 25 amperes. Can 16 general-use receptacles in an office building be connected to a 20-ampere circuit?

   Answer: _____

   _____

   _____

   _____

   Reference: _____

10. When an electric heat pump is installed with backup resistance heat in a building, (a) could one of the loads ever be considered dissimilar for the purposes of calculating the service loads? (b) Would the heat pump load have to be calculated at 125 percent for the service sizing?

    Answer: _____

    _____

    _____

    _____

    Reference: _____

11. Are all 20-ampere residential underground circuits required to have GFCI protection?

    Answer: _____

    _____

    _____

    _____

    Reference: _____

12. Can two circuit breakers mounted adjacent to each other in a panelboard with handle ties be used in place of a double-pole breaker to supply a 240-volt electric baseboard heater?

Answer: _____

_____

_____

_____

Reference: _____

13. Are water heaters considered to be a continuous load so that a 125 percent load is required for calculating feeder and service on an installation?

Answer: _____

_____

_____

_____

Reference: _____

14. Is it permitted to plug a microwave oven with a nameplate rating of 13 amperes into a 15-ampere receptacle protected on a 20-ampere circuit?

Answer: _____

_____

_____

_____

Reference: _____

15. Can a 1400 VA load, a cord-and-plug-connected load, be supplied by a 240-volt circuit in a residence?

Answer: _____

_____

_____

_____

Reference: _____

16. Three receptacles on a single yoke or strap are to be installed in a commercial or industrial facility. Are these outlets to be calculated at 180 VA or 540 VA?

Answer: _____

_____

_____

_____

Reference: _____

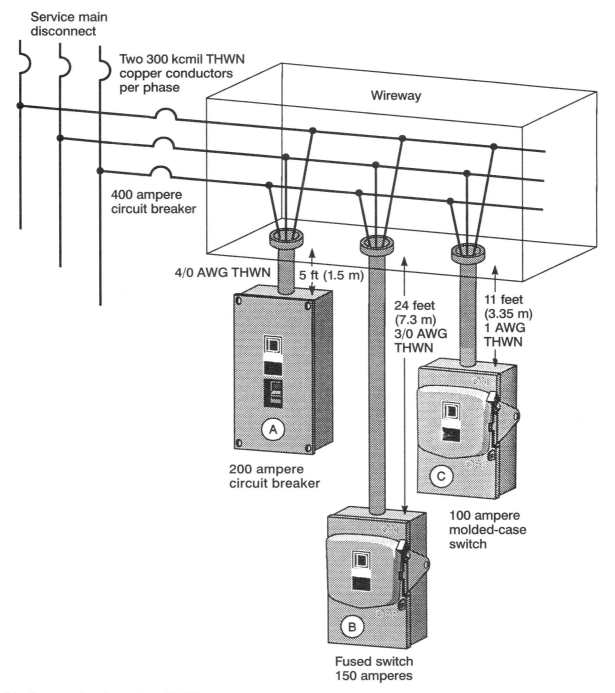

Service main disconnect

Two 300 kcmil THWN copper conductors per phase

Wireway

400 ampere circuit breaker

4/0 AWG THWN

5 ft (1.5 m)

200 ampere circuit breaker

A

24 feet (7.3 m) 3/0 AWG THWN

11 feet (3.35 m) 1 AWG THWN

100 ampere molded-case switch

C

Fused switch 150 amperes

B

Use this diagram for Questions 17–19.

17. Does the circuit breaker marked Ⓐ comply with the *NEC*®? If so, which section?

Answer: _____

_____

_____

_____

Reference: _____

18. Does the switch marked Ⓑ comply with the *NEC*? If so, which section?

Answer: _____

_____

_____

_____

Reference: _____

19. Does the molded-case switch marked Ⓒ comply with the *NEC*?

Answer: _____

_____

_____

_____

Reference: _____

20. *NEC* requires overcurrent protection where the busway is reduced in size for the last 50 feet (15.24 m) or more in other than industrial installations. True or False?

Answer: _____

_____

_____

_____

Reference: _____

Chapter Nine

GROUNDING

NEC® Article 250 covers the general requirements and many specific requirements for grounding such as systems, circuits, and equipment required, permitted, and/or not permitted to be grounded, circuit conductors to be grounded on grounded systems, location of the grounding connections, type and sizes of grounding and bonding conductors and electrodes, the methods of grounding and bonding, and the conditions under which guards, isolation, or insulation can substitute for grounding. Other applicable articles applying to particular cases of installations of conductors and equipment are found in *NEC®* 250.4. (See Figure 9–1.)

The rules and performance requirements of grounding and bonding are clearly stated in 250.4(A) and (B). Article 250 has been formatted to the proper *NEC®* style and renumbered. The majority of exceptions and fine print notes have been rewritten into positive language or deleted. The specific requirements have been grouped into A and B. A is titled as follows:

- (A)(1) Electrical System Grounding
- (A)(2) Grounding of Electrical Equipment
- (A)(3) Bonding of Electrical Equipment
- (A)(4) Bonding of Electrically Conductive Materials and Other Equipment
- (A)(5) Effective Ground-Fault Current Path

NEC® Part VIII for direct-current systems, Part II for alternating-current circuits and systems (see Figure 9–2), and 250.34 for portable and vehicle mounted generators outline the conditions for which these systems and circuits are to be grounded or can be grounded. *NEC®* 250.22 covers the circuits not to be grounded. *NEC®* 250.24 and 250.32 cover the requirements necessary for grounding buildings, single or locations with two or more buildings fed from a common service.

Example: The following example illustrates how the size of the grounded (neutral) wire may be selected. A new 3-phase, 4-wire feeder is connected in an existing junction box to three single-phase plus grounded circuits, each consisting of two-phase conductors and a grounded conductor. Two of these circuits are in one conduit; the third is in another. The single-phase circuit conductors are all 2 AWG THW copper. Determine the ampacity of each single-phase circuit and the required size of new service if electric discharge lighting ballasts are in the circuits, but the load is considered noncontinuous.

Step 1. Neutrals in these circuits must be counted as current-carrying conductors. (*NEC®* 310.15(B)(4).) The ampacity of 2 AWG THW conductors from Table 310.16 is 115 amperes.

Step 2. The circuit ampacity in the conduit with the single circuit; derating is not required for only three conductors. (*NEC®* Table 310.15(B)(2)(A).) Thus, the allowable ampacity is 115 amperes.

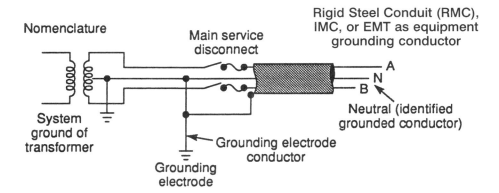

Requirements for grounding electrode conductor, grounded system

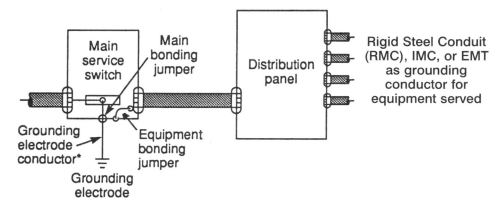

*Grounds both the identified grounded conductor and the equipment grounding conductor. It is connected to the grounded metalic cold water piping on premises, or other available grounding electrode, such as metal building frame if effectively grounded, or a made electrode.

Figure 9–1 Grounding principles. *(Courtesy of American Iron & Steel Institute)*

Step 3. In the conduit with two circuits, an 80 percent derating factor must be used, because all six wires are considered as current-carrying conductors (*NEC®* Table 310.15(B)(2).) Each of these two circuits has a design ampacity of $80\% \times 115 = 92$.

Step 4. The sum of the ampacities of all three circuits is the required ampacity of the service conductors and the service neutral (220.61).

Step 5. The value is $115 + 92 + 92 = 299$. Each conductor and neutral requires not less than a 350-kcmil 75°C conductor ampacity. Reference Article 310, Table 310.16

NEC® 250.30 outlines the requirements for grounding a separately derived alternating-current system. Article 250, Part IV covers the requirements for grounding the enclosures, Part V the equipment grounding requirements, Part VI covers the methods for grounding, and Part VII the bonding requirements. (See Figure 9–3 and Figure 9–4.)

Part V clearly delineates between service equipment and other-than-service equipment.

Caution: It is extremely important for only one grounding system to exist in an installation because multiple-separated grounds can create unwanted ground loops that may create errors in data transmission and malfunctions. Where a ground loop exists, known techniques will solve the problem. Although the single-point ground is preferred, it may not always be feasible where long distances and very high frequencies are present.

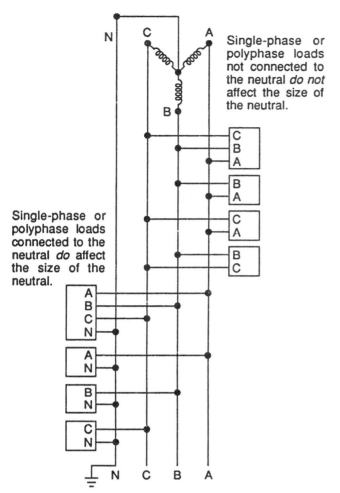

Figure 9–2 Size of a neutral conductor as in *NEC®* 220.61. *(Courtesy of American Iron & Steel Institute)*

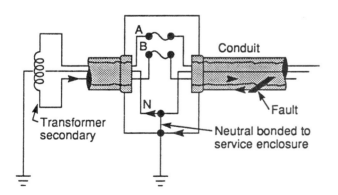

Figure 9–3 Bonding at service disconnect enclosure as required in *NEC®* 250.92(A) and (B). *(Courtesy of American Iron & Steel Institute)*

Part III clearly describes the grounding electrode systems (see Figure 9–4), and the grounding conductors, types, sizes, and enclosures for those conductors. Most important are Table 250.66 for sizing the

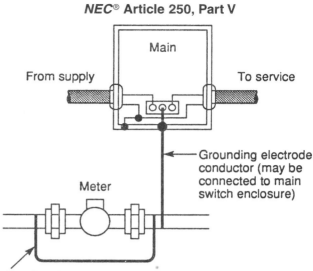

NEC® **Article 250, Part V**

Bonding jumper around water meter and other fittings in water piping likely to be disconnected, such as unions, where grounding connection is not on street side of water meter.

Figure 9–4 Bonding jumpers as in *NEC®* Article 250, Part V. *(Courtesy of American Iron & Steel Institute)*

grounding electrode conductor for an ac system, and 250.66(A) through (C) as they apply to that table. Table 250.122 lists the minimum equipment grounding conductor for grounding raceway and equipment. *NEC®* 250.122 provides rules for applying that table. Part IX covers instrument transformers, and so on. Part X covers systems and circuits of 1 kV (1000 volts) or over (high voltage). Where high-voltage systems are grounded, they must comply with all applicable provisions of all Article 250, Part X, which supplements or modifies Parts I through IX. *NEC®* 250.80 and 250.86 clarify a common misconception of adding an insulated equipment grounding conductor with the circuit conductors from the source, service, or subpanel to the equipment being served and terminating on the ends relieves one responsibility of installing the metallic raceways as a complete system. All metal raceway enclosures must be bonded to the equipment grounding conductor to ensure that no potential exists on that equipment, and if one of the conductors fault to the raceway would occur, that the overcurrent device will trip. The insulated conductor when installed is redundant and not required. (The redundant conductor will only carry about 3 percent to 5 percent of the current flowing, while the metal raceway will carry 95 percent to 97 percent of the current flowing.) Article 250 describes the general requirements for equipment grounding and provides a list of specific items, both the electrical and nonelectrical equipment fastened in place, and cord-connected equipment in residential occupancies and other than residential occupancies. The list is specific. Article 250 also covers the limited requirement for spacing lightning rods, 250.106 supplements the requirements in NFPA 780, which contains detailed information on grounding lightning protection systems. The types of grounding-electrodes and details for the grounding electrode system are found in 250.50 through 250.60. Related to those sections, you also should read 250.104, which is the bonding requirements for piping systems and exposed building steel, (a) covering metal water piping and (b) other metal piping, which provide additional safety to the system. As you read 250.50 and 250.52, you will see that all available grounding electrodes on each premise must be used and bonded together. One electrode system must be provided where the building contains others, such as building steel, metal water pipes, and a concrete-encased electrode; all must be used and bonded together. Read these sections carefully before applying them. *NEC*® 250.58 requires common grounding electrodes and states:

Where an ac system is connected to a grounding electrode in or at a building as specified in 250.24 and 250.32, the same electrode shall be used to ground conductor enclosures and equipment in or on that building. Where separate services supply a building and are required to be connected to a grounding electrode, the same grounding electrode shall be used.

*Two or more grounding electrodes that are effectively bonded together shall be considered as a single grounding electrode system in this sense.**

Example: A dairy barn is fed underground from the farmhouse with a 225-ampere overcurrent device using 4/0 AWG copper phase conductors and a 1 AWG (neutral) grounded conductor. What size equipment grounding conductor is required for this installation?

Step 1. Index "Grounding, Separate buildings," 250.32. Also, "Grounding Conductors, Sizes," Table 250.122.

Step 2. *NEC*® 250.32 requires an insulated or covered equipment grounding conductor where livestock is housed. Also, 547.9 (B)(1) contains similar requirements.

Step 3. Table 250.122. For the minimum size permitted for a 225-ampere overcurrent

device, the table lists only a 200 and a 300. Therefore, the 300 would apply, because the 200 would not be adequate.

Answer: 4 AWG copper

When sizing the equipment grounding conductor in accordance with Table 250.122, beware that this is the minimum size conductor allowed but may not be large enough to comply with other applicable sections in Article 250, such as 250.4(A)(5), which requires an effective ground-fault current path: *Electrical equipment and wiring and other electrically conductive material likely to become energized shall be installed in a manner that creates a permanent, low impedance circuit capable of safely carrying the maximum ground-fault current likely to be imposed on it from any point on the wiring system where a ground fault may occur to the electrical supply source. The earth shall not be used as the sole equipment grounding conductor or fault current path.** (This requirement clarifies that driving a grounding electrode at remote equipment is not an acceptable means for grounding.) An equipment grounding conductor must be run with the other conductors as stated in 300.3(B) from the supply to the load and sized in accordance with Table 250.122 or larger to facilitate the operation of the overcurrent device protecting that circuit. You will find after studying Article 250 that grounding is one of the most interesting sections of the *NEC*®, and it is written for easy understanding. However, careful study and application are necessary to accomplish a safe design and/or installation of the electrical system.

Example: The following example illustrates how the size of the grounded (neutral) wire may be selected. A new 3-phase, 4-wire feeder is connected in an existing junction box to three single-phase plus grounded circuits, each consisting of two phase conductors and a grounded conductor. Two of these circuits are in one conduit, the third is in another. The single-phase circuit conductors are all 2 AWG THW copper. Determine the ampacity of each single-phase circuit and the required size of new service if electric discharge lighting ballasts are in the circuits, but the load is considered noncontinuous.

Solution: Neutrals in these circuits must be counted as current-carrying conductors (*NEC*® 310.15(B)(4)(a)). The ampacity of 2 AWG THW conductors from Table 310.16 is 115 amperes. The circuit ampacity in the conduit with the single circuit, and no derating is required for only three conductors. Thus, the ampacity equals 115 amperes. In the conduit with two circuits, an 80 percent derating factor must be used, because all six wires are considered as current-carrying conductors. Each of these two circuits has a design ampacity of 80% × 115 = 92. The sum of the ampacities of three circuits is the required ampacity of the service conductors and the service neutral (220.22). This value is 115 + 92 + 92 = 299. Each conductor and neutral requires not less than a 350-kcmil 75 percent seat conductor ampacity. Reference Article 310, Table 310.16.

Computer Model Developed

It is with a great deal of satisfaction that I can report that the excellent work that has carried this industry for the past forty years, developed and written by

R. H. "Dick" Kaufman (General Electric "GER 957A," *Some Fundamentals of Equipment Grounding Circuit Design,* IE 1058.33 November 1954, Applications and Industry, Vol. 73, Part II) and

Eustace C. Soares (Pringle Switch, *Grounding Electrical Distribution Systems for Safety,* 1966 through the 5th Edition *IAEI Soares Book on Grounding,* 1993, by J. Philip Simmons)

has been revisited.

In early 1993 at the request of J. Philip Simmons, author of the 4th and 5th editions of *IAEI Soares Book on Grounding*, the steel conduit and tubing producer members of the National Electrical Manufacturers Association (NEMA) made funding available to Georgia Institute of Technology's School of Electrical and Computer Engineering to develop a computer model. A. P. Sakis Meliopoulos, P.E., and Elias N. Glytsis, P.E., developed a model that was validated by field tests consisting of arc-voltage testing, and fault-current testing on thirteen 256-foot runs of electrical metallic tubing with both steel and die-cast set-screw and compression fittings, rigid steel conduit and intermediate steel conduit with conductors enclosed at Kearney Laboratories at McCook, Illinois. Mr. Meliopoulos stated in his final report that the techniques established by *Soares Book on Grounding* has served this industry well in providing primary design guidance. (See Figure 9–5, Figure 9–6, Figure 9–7, Figure 9–8, Figure 9–9, and Figure 9–10.)

Fortunately, computer and analytical means, which have been developed at a rapid pace, have now made it possible to develop and validate a model utilizing current data that can be used with confidence.

This new data will replace the only data available to date. Today's manufacturing processes have changed both the quality of material used to produce electrical materials and the consistency of those products. Although hand calculations were fairly accurate to evaluate given criteria, this new model will offer a broad band of conditions and the user no longer has to extrapolate data. Mr. Meliopoulos and I presented a paper through IAS/IEEE in 1997 at the Fall Meeting, and the model is now available. The attachments to this report are excerpts from the final report. It is interesting to note that the tests and research of new materials and circuit-breaker test guidelines revealed that the criteria of 50 arc volt and 500 percent (5IP) of overcurrent device rating were overly conservative, and he recommended an arc voltage of 40 and a 4IP (4 times the rating of the overcurrent device, i.e., 30-ampere circuit-breaker, 4IP = 30 × 400% = 120) of the overcurrent device, and as stated these are still conservative numbers.

"Modeling and Testing of Steel EMT, IMC and Rigid (GRC) Conduit." Prepared by A. P. Sakis Meliopoulos and Elias N. Glytsis, School of Electrical and Computer Engineering, Georgia Institute of Technology, Atlanta, GA 30332.

For copies of this report and the free software, contact: The Steel Tube Institute of North America, http://www.steelconduit.org.

Even when the raceway does not provide the sole equipment ground but is supplemented by an equipment grounding conductor wire or a bus enclosed within, the raceway provides a parallel path and also protects the wire and bus from physical damage. Where connected as a parallel path, the raceway not only lowers the total impedance of the grounding path but carries 95 percent or more of the current. Furthermore, the raceway provides additional protection as a grounding shield; for instance, in the unlikely event that a phase conductor becomes damaged within a raceway and the conductor is exposed and comes in contact with a grounded raceway, the resulting fault current would be carried to the grounded service equipment enclosure through the main bonding jumper to the system and the overcurrent device, thus clearing the fault.

(Text continues on page 130)

Examples of Maximum Length Equipment Grounding Conductor (Steel EMT, IMC, GRC, and Copper or Aluminum Wire) Calculated as a Safe Return Fault Path to Overcurrent Device Based on 1997 Georgia Tech Software Version (GEMI 2.4, 1998) With an Arc Voltage of 40 and 4 IP at 25°C Ambient Based on a Circuit Voltage of 120 Volts to Ground THHN/THWN Insulation

Overcurrent Device Rating Amperes (75°C)	400% (4IP) Overcurrent Device Rating Amperes	Circuit Conductor Size AWG/kcmil Copper or Aluminum	Steel EMT, IMC, GRC Trade Size	(1) Equipment Grounding Conductor Size AWG Copper or Aluminum	Length of EMT Run Calculated Maximum (In Feet)	Length of IMC Run Calculated Maximum (In Feet)	Length of GRC Run Calculated Maximum (In Feet)	Copper Grounding Conductor w/o Steel Conduit Maximum Run (In Feet)	Aluminum Grounding Conductor w/o Steel Conduit Maximum Run (In Feet)
20	80	12	1/2 (16)	—	395	398	384	—	—
20	80	12	—	12	—	—	—	300	—
20	80	10 AL	—	10 AL	—	—	—	—	293
30	120	10	1/2 (16)	—	358	383	364	—	—
30	120	10	3/4 (21)	—	404	399	386	—	—
30	120	10	—	10	—	—	—	319	—
30	120	8 AL	—	8 AL	—	—	—	—	310
40	160	8	3/4 (21)	—	407	414	395	—	—
40	160	8	1 (27)	—	447	431	418	—	—
40	160	8	—	10	—	—	—	294	—
40	160	8 AL	—	8 AL	—	—	—	—	232
60	240	6	3/4 (21)	—	350	383	363	—	—
60	240	6	1 (27)	—	404	400	382	—	—
60	240	6	—	10	—	—	—	228	—
60	240	4 AL	—	8 AL	—	—	—	—	221
100	400	3	1 1/4 (35)	—	402	397 (4)	373	—	—
100	400	3	—	8	—	—	—	229	—
100	400	1 AL	—	6 AL	—	—	—	—	222
200	800	3/0	2 (53)	—	390	389	363	—	—
200	800	3/0	—	6	—	—	—	201	—
200	800	250 AL	—	4 AL	—	—	—	—	195

Note: Software is not limited to above examples.

(1) Per 1999 NEC® Table 251-18

Figure 9–5 (Table derived from Software [SCA] and Testing developed at Georgia Institute of Technology and sponsored by the producers of Steel EMT, IMC, and rigid steel conduit.)

Examples of Maximum Length Equipment Grounding Conductor (Steel EMT, IMC, GRC, and Copper or Aluminum Wire) Calculated as a Safe Return Fault Path to Overcurrent Device Based on 1997 Georgia Tech Software Version (GEMI 2.4, 1998) With an Arc Voltage of 40 and 4 IP at 25°C Ambient Based on a Circuit Voltage of 277 Volts to Ground THHN/THWN Insulation

Overcurrent Device Rating Amperes (75°C)	400% (4IP) Overcurrent Device Rating Amperes	Circuit Conductor Size AWG/kcmil Copper or Aluminum	Steel EMT, IMC, GRC Trade Size	(1) Equipment Grounding Conductor Size AWG Copper or Aluminum	Length of EMT Run Calculated Maximum (In Feet)	Length of IMC Run Calculated Maximum (In Feet)	Length of GRC Run Calculated Maximum (In Feet)	Copper Grounding Conductor w/o Steel Conduit Maximum Run (In Feet)	Aluminum Grounding Conductor w/o Steel Conduit Maximum Run (In Feet)
20	80	12	1/2 (16)	—	1170	1179	1140	—	—
20	80	12	—	12	—	—	—	890	—
20	80	10 AL	—	10 AL	—	—	—	—	870
30	120	10	1/2 (16)	—	—	1135 (4)	1143	—	—
30	120	10	3/4 (21)	—	1199	1182	—	—	—
30	120	10	—	10	—	—	—	946	—
30	120	8 AL	—	8 AL	—	—	—	—	920
40	160	8	3/4 (21)	—	1208	1228	1170	—	—
40	160	8	1 (27)	—	1326	1276	1239	—	—
40	160	8	—	10	—	—	—	871	—
40	160	8 AL	—	8 AL	—	—	—	—	690
60	240	6	3/4 (21)	—	1039	1134	1075	—	—
60	240	6	1 (27)	—	1197	1186	1131	—	—
60	240	6	—	10	—	—	—	676	—
60	240	4 AL	—	8 AL	—	—	—	—	657
100	400	3	1 1/4 (35)	—	1192	1176 (4)	1107	—	—
100	400	3	—	8	—	—	—	680	—
100	400	1 AL	—	6 AL	—	—	—	—	659
200	800	3/0	2 (53)	—	1157	1155	1077	—	—
200	800	3/0	—	6	—	—	—	598	—
200	800	250 AL	—	4 AL	—	—	—	—	578

Note: Software is not limited to above examples.

(1) Per 1999 *NEC*® Table 251-18

Figure 9–6 (Table derived from Software [SCA] and Testing developed at Georgia Institute of Technology and sponsored by the producers of Steel EMT, IMC, and rigid steel conduit.)

Maximum Length of Electrical Metallic Tubing That May Safely Be Used as an Equipment-Grounding Circuit Conductor.
Based on a Ground-Fault Current of 400% of the Overcurrent Device Rating.
Circuit 120 Volts to Ground; 40 Volts Drop at the Point of Fault.
Ambient Temperature 25°C

EMT Trade Size	Conductors AWG No.	Overcurrent Device Rating Amperes 75°C*	Fault Clearing Current 400% O.C. Device Rating Amperes	Maximum Length of EMT Run in Feet
1/2 (16)	3-12	20	80	395
	4-10	30	120	358
3/4 (21)	4-10	30	120	404
	4-8	50	200	332
1 (27)	4-8	50	200	370
	3-4	85	340	365
1 1/4 (35)	3-2	115	460	391
1 1/2 (41)	3-1	130	520	407
	3-2/0	175	700	364
2 (53)	3-3/0	200	800	390
	3-4/0	230	920	367
2 1/2 (63)	3-250 kcm	255	1020	406
3 (78)	3-350 kcm	310	1240	404
	3-500 kcm	380	1520	370
	3-600 kcm	420	1680	353
4 (103)	3-900 kcm	520	2080	353
	3-1000 kcm	545	2180	347

* 60°C for 20- and 30-ampere devices.

Based on 1994 Georgia Tech Model

Figure 9–7 (Table derived from Software [SCA] and Testing developed at Georgia Institute of Technology and sponsored by the producers of Steel EMT, IMC, and rigid steel conduit.)

Maximum Length of Intermediate Metal Conduit That May Safely Be Used as an Equipment-Grounding Circuit Conductor.
Based on a Ground-Fault Current of 400% of the Overcurrent Device Rating.
Circuit 120 Volts to Ground; 40 Volts Drop at the Point of Fault.
Ambient Temperature 25°C

IMC Trade Size	Conductors AWG No.	Overcurrent Device Rating Amperes 75°C*	Fault Clearing Current 400% O.C. Device Rating Amperes	Maximum Length of IMC Run in Feet
1/2 (16)	3-12	20	80	398
	4-10	30	120	383
3/4 (21)	4-10	30	120	399
	4-8	50	200	350
1 (27)	4-8	50	200	362
	3-4	85	340	382
1 1/4 (35)	3-2	115	460	392
1 1/2 (41)	3-1	130	520	402
	3-2/0	175	700	377
2 (53)	3-3/0	200	800	389
	3-4/0	230	920	375
2 1/2 (63)	3-250 kcm	255	1020	368
3 (78)	3-350 kcm	310	1240	367
	3-500 kcm	380	1520	338
	3-600 kcm	420	1680	325
4 (103)	3-900 kcm	520	2080	320
	3-1000 kcm	545	2180	314

* 60°C for 20- and 30-ampere devices.

Based on 1994 Georgia Tech Model

Figure 9–8 (Table derived from Software [SCA] and Testing developed at Georgia Institute of Technology and sponsored by the producers of Steel EMT, IMC, and rigid steel conduit.)

Maximum Length of Galvanized Rigid Conduit That May Safely Be Used as an Equipment-Grounding Circuit Conductor.
Based on a Ground-Fault Current of 400% of the Overcurrent Device Rating.
Circuit 120 Volts to Ground; 40 Volts Drop at the Point of Fault.
Ambient Temperature 25°C

GRC Trade Size	Conductors AWG No.	Overcurrent Device Rating Amperes 75°C*	Fault Clearing Current 400% O.C. Device Rating Amperes	Maximum Length of GRC Run in Feet
1/2 (16)	3-12	20	80	384
	4-10	30	120	364
3/4 (21)	4-10	30	120	386
	4-8	50	200	334
1 (27)	4-8	50	200	350
	3-4	85	340	357
1 1/4 (35)	3-2	115	460	365
1 1/2 (41)	3-1	130	520	377
	3-2/0	175	700	348
2 (53)	3-3/0	200	800	363
	3-4/0	230	920	347
2 1/2 (63)	3-250 kcm	255	1020	356
3 (78)	3-350 kcm	310	1240	355
	3-500 kcm	380	1520	327
	3-600 kcm	420	1680	314
4 (103)	3-900 kcm	520	2080	310
	3-1000 kcm	545	2180	304

* 60°C for 20- and 30-ampere devices.

Based on 1994 Georgia Tech Model

Figure 9–9 (Table derived from Software (SCA) and Testing developed at Georgia Institute of Technology and sponsored by the producers of Steel EMT, IMC, and rigid steel conduit.)

Maximum Length of Equipment-Grounding Conductor That May Safely Be Used as an Equipment-Grounding Circuit Conductor. Based on a Ground-Fault Current of 400% of the Overcurrent Device Rating. Circuit 120 Volts to Ground; 40 Volts Drop at the Point of Fault. Ambient Temperature 25°C

Copper Equipment Grounding Conductor AWG Size***	Copper Circuit AWG Conductors	Maximum Length of Run (in feet) using Copper Equipment Ground Conductor	Aluminum Equipment Grounding Conductor AWG Size***	Aluminum Circuit AWG Conductors	Maximum Length of Run (in feet) using Aluminum Equipment Ground Conductor	Overcurrent Device Rating Amperes 75°C**	Fault Clearing Current 400% O.C. Device Rating Amperes
14	14	253	12	12	244	15	60
12	12	300	10	12	226	20	80
10	10	319	8	8	310	30	120
10	8	294	8	8	232	40	160
10	6	228	8	4	221	60	240
8	3	229	6	1	222	100	400
6	3/0	201	4	250 kcm	195	200	800
4	350 kcm	210	2	500 kcm	204	300	1200
3	600 kcm	195	1	900 kcm	192	400	1600
2	2-4/0	160	1/0	2-400 kcm	163	500	2000
1	2-300 kcm	160	2/0	2-500 kcm	161	600	2400
1/0	3-300 kcm	134	3/0	3-400 kcm	131	800	3200
2/0	4-250 kcm	114	4/0	4-400 kcm	115	1000	4000
3/0	4-300 kcm	106	250 kcm	4-500 kcm	107	1200	4800
4/0	4-600 kcm	93	350 kcm	4-900 kcm	97	1600	6400
250 kcm	5-600 kcm	78	400 kcm	5-800 kcm	79	2000	8000
350 kcm	6-600 kcm	*	600 kcm	6-900 kcm	*	2500	10,000
400 kcm	8-500 kcm	*	600 kcm	8-750 kcm	*	3000	15,000
500 kcm	8-1000 kcm	*	800 kcm	8-1500 kcm	*	4000	16,000
700 kcm	10-1000 kcm	*	1200 kcm	10-1500 kcm	*	5000	20,000
800 kcm	12-1000 kcm	*	1200 kcm	12-1500 kcm	*	6000	24,000

Based on 1994 Georgia Tech Model

* Calculations necessary
** 60°C for 20- and 30-ampere devices
*** Based on *NEC*® Chapter 9 Table 8

Figure 9–10   (Table derived from Software (SCA) and Testing developed at Georgia Institute of Technology and sponsored by the producers of Steel EMT, IMC, and rigid steel conduit.)

By contrast, a damaged phase conductor run as open wiring or in a nonmetallic raceway could remain exposed to accidental contact indefinitely. A fault within a nonmetallic raceway or open runs of conductors may or may not provide a low impedance return path to trip the overcurrent device protecting the circuit.

An equipment grounding conductor and a grounded circuit conductor are not the same electrically. It is unsafe to connect the grounded circuit conductor to the raceway or to any other noncurrent-carrying equipment at any point on the load side of the service equipment. The connections permitted to be made on the line and load sides of the service disconnecting equipment are specified in 250.142 of the *NEC*®, Parts I and II, which clarifies the preferred design method for location of these connections.

In 1998, the electronic and circuit-breaker industry introduced a new level of safety to the *NEC*®, the arc-fault circuit-breaker. Several proposals (2-128, 2-129, and 2-130) were submitted to add this new level of safety device designed to protect dwellings. The substantiation included U.S. Consumer Product Safety Commission (CPSC) fire records. They stated that there were over 40,000 fires resulting in approximately 250 deaths annually and that these figures had not decreased over the past decade. Based on a survey by a large insurance company, approximately 36 percent of these fires originated in permanent wiring, 50 percent in appliances, and 12 percent in cords. They analyzed these reports and stated that approximately 38 percent were from arcing faults and about 12 percent were from loose connections.

The solution was to add a listed arc-fault circuit interrupter (AFCI) device. The AFCI device has electronic circuitry capable of recognizing the unique current and/or voltage signatures of an arcing fault. When the arc fault is detected, it acts similarly to the GFCI device and interrupts the circuit. This technology integrates a conventional thermal circuit-breaker with an AFCI.

The *National Electrial Code*® has defined this device in 210.12. The requirement is that all new dwelling bedroom circuits be an AFCI type.

QUESTION REVIEW

Each lesson is designed purposely to require the student to apply the entire *NEC®* text, and not specific chapters or articles. It has been found that when studying for a timed, open-book examination, the student must gain proficiency in the Table of Contents, the Index, and the ability to move quickly from cover to cover to find the correct answer to each question in a timely fashion.

1. Required grounding conductors and bonding jumpers cannot be connected solely by _____ connections.

   Answer: _____

   _____

   _____

   _____

   Reference: _____

2. The conductor permitted to bond together all isolated noncurrent-carrying metal parts of an outline lighting system is at least size _____ AWG copper.

   Answer: _____

   _____

   _____

   _____

   Reference: _____

3. A connection to a concrete-encased, driven or buried grounding electrode shall be _____.

   Answer: _____

   _____

   _____

   _____

   Reference: _____

4. The grounded conductor of an electrical branch-circuit is identified by the color _____ .

   Answer: _____

   _____

   _____

   _____

   Reference: _____

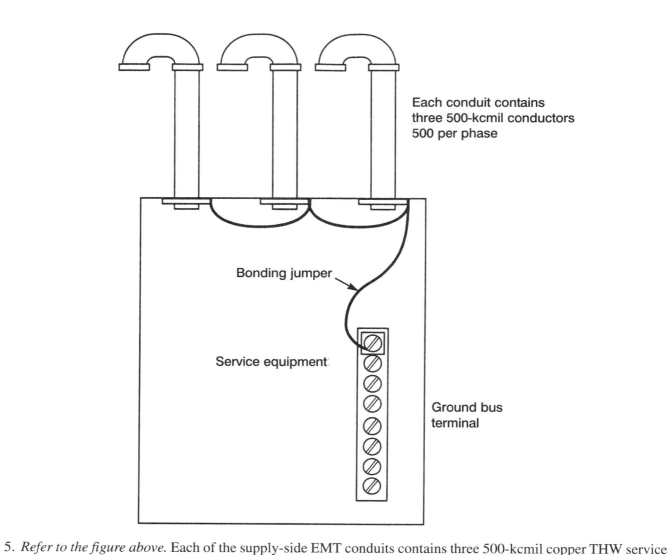

Each conduit contains
three 500-kcmil conductors
500 per phase

Bonding jumper

Service equipment

Ground bus
terminal

5. *Refer to the figure above.* Each of the supply-side EMT conduits contains three 500-kcmil copper THW service conductors in parallel. As shown, only one bonding jumper is used to bond all the conduits to the grounded bus terminal. This bonding jumper must be at least size _____ AWG copper.

Answer: _____

_____

_____

_____

Reference: _____

6. A rigid metal conduit contains three circuits: two 150-ampere, 3-phase circuits, and one 300-ampere, single-phase circuit. The load side service equipment bonding jumper for this conduit must be at least size _____ AWG copper.

Answer: _____

_____

_____

_____

Reference: _____

7. Are EMT set-screw connectors approved for grounding of service raceways or are jumpers required to bond around them?

   Answer: _____

   _____

   _____

   _____

   Reference: _____

8. Where feeder conductors are paralleled and routed in separate nonmetallic raceways, is an equipment grounding conductor required in each nonmetallic raceway or only in one nonmetallic raceway?

   Answer: _____

   _____

   _____

   _____

   Reference: _____

9. Are submersible deep-well pumps required to have an equipment grounding conductor run with the other conductors and the submersible pump motor housing grounded?

   Answer: _____

   _____

   _____

   _____

   Reference: _____

10. Is it permissible to ground the secondary of a separately derived system to the grounded terminal in the service switchboard instead of the nearest effectively grounded structural steel or nearest water pipe?

    Answer: _____

    _____

    _____

    _____

    Reference: _____

11. Where installing the grounding-electrode conductor in a raceway for physical protection, is the metal raceway required to be bonded to the grounding-electrode conductor? If the answer is yes, is it sufficient to bond it at the service panel only or must bonding be done at all terminations?

    Answer: _____

    _____

    _____

    _____

    Reference: _____

12. Where installing the electrical hookups for RV sites in a recreational vehicle park, can a ground rod be driven at each RV site instead of pulling an equipment grounding conductor with a circuit conductor from the source?

    Answer: _____

    _____

    _____

    _____

    Reference: _____

13. You have recently installed a separately derived system, a dry-type transformer in a commercial building. No effectively grounded structural steel was near the transformer location. The transformer secondary is grounded to a nearby metal water pipe. Is a supplemental grounding electrode required?

    Answer: _____

    _____

    _____

    _____

    Reference: _____

14. When bonding the noncurrent-carrying metal parts of a recreational vehicle, what is the minimum size copper or equivalent bonding conductor?

    Answer: _____

    _____

    _____

    _____

    Reference: _____

15. Does the *Code* require interior gas piping systems to be bonded to the grounding-electrode system?

    Answer: _____

    _____

    _____

    _____

    Reference: _____

16. Two parallel runs of four 500-kcmil conductors each are run in rigid nonmetallic conduits to feed a 600-ampere panel. Is an equipment grounding conductor required in each conduit? What size equipment grounding conductor is required in each conduit?

    Answer: _____

    _____

    _____

    _____

    Reference: _____

17. When terminating the equipment grounding conductor in an outlet box supplied by nonmetallic-sheathed cables, can the equipment grounding conductors be twisted together or must they be connected with a wire connector?

Answer: _____

_____

_____

_____

Reference: _____

18. What size aluminum equipment grounding conductor will be required to be run with the circuit conductors fed from a 60-ampere fusible switch?

Answer: _____

_____

_____

_____

Reference: _____

19. In a run of conduit supplying a motor, it is necessary to install 4 feet (1.2 m) of flexible metal conduit where the raceway leaves the panel to get around a column. When the conduit reaches the motor, an additional 3 feet (900 mm) of flexible metal conduit is employed for convenience and flexibility. Is an equipment grounding conductor required in this metal raceway?

Answer: _____

_____

_____

_____

Reference: _____

20. When making an installation of a high-impedance grounded neutral system, how is the neutral conductor to be sized?

Answer: _____

_____

_____

_____

Reference: _____

Chapter Ten

Studying this chapter along with the 2005 *NEC®* gives the reader a basic introduction to the branch-circuit and feeder requirements in the *Code* that apply to all installations for above and below 600 volts.

After studying this chapter, you should know:

General Wiring Methods

- The definitions and the application of terms that are applicable and that are unique to the 2005 *NEC®*
- The identical section numbering scheme now used in Chapter 3
- The new use of acronyms applying to raceways and cables
- The importance of proper selection of the wiring method raceway type or cable type for the application

Cable Types

- The new *Code* articles in which cables are addressed as they apply to the 2005 *NEC®*
- The unique requirements for individual metallic sheathed cable types
- The unique requirements for individual nonmetallic sheathed cable types
- The common requirements for all cable types

Raceway Types

- The unique requirements for individual raceway types
- The common requirements for metal types
- The common requirements for nonmetallic types
- The calculation methods for determining expansion
- The calculation methods for determining the conductor fill and raceway sizing

Enclosures, Panelboards, Switchboards, and Switches

- The enclosures and equipment applicatons as they apply to the 2005 *NEC®*

Cable Tray Systems

- The fundamentals for applying the requirements to cable tray systems as applicable to the 2005 *NEC®*

WIRING METHODS

The general requirements for wiring methods are found in Article 300. (See Figure 10–1, Figure 10–2, and Figure 10–3.) As you may recall in earlier chapters of this book, several times you were told that you must know the contents of Article 300 thoroughly because it is used in some application for every installation.

The general wiring methods in Article 300 include requirements that apply to the specific wiring methods, such as nonmetallic-sheathed cable, which is covered in Article 334 (see Figure 10–4), as specific support requirements and installation criteria, such as found in 300.4(A), (B), (C), (D), (E), and (F) and the Exceptions. These requirements are mandatory and must be adhered to when making an installation of any wiring method and are in Chapter 3 as they apply. Chapter 3 also covers burial depths (in Table 300.5 for conductors under 600 volts and Table 300.50 for conductors over 600 volts) and many conditions that apply to underground installations, such as splices, taps, and the backfill material used so as not to damage the raceway or cable that you have selected as a wiring method for your installation. When a wiring method is installed in earth that is not suitable as physical protection, select fill material should be used, such as sand, or soil without rocks. This article also covers such things as seals and protection against corrosion, indoor wet locations, raceways exposed to different temperatures, and expansion joints. This chapter discusses much about the expansion joint requirements for nonmetallic raceways. However, 300.7 covers all wiring methods, and where they are required to compensate for expansion or contraction. (See Figure 10–5, Figure 10–6, and Figure 10-7.) The electrical and mechanical continuity for metallic systems, both the raceway and the enclosures, *NEC®* 300.11, requires that these methods, the raceways, cable assemblies, boxes, cabinets, and fittings, all be securely fastened where supported on support wires above suspended ceilings (refer to Figure 10–4).

For branch-circuit wiring, read this section very carefully. The wiring systems for feeders or services

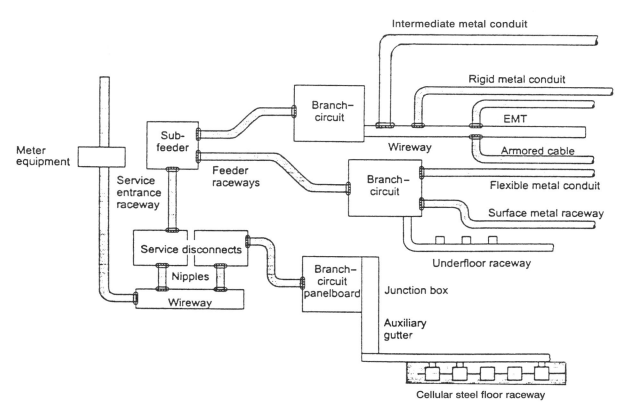

Figure 10–1 Service raceway per *NEC*® Article 230. ***Note:*** Location of overcurrent devices not shown. *(Courtesy of American Iron & Steel Institute)*

are not permitted to be secured in this manner; only certain branch-circuits are permitted. Specifically, the junction box out of the box section in Article 314 places more stringent requirements for supporting these box systems. All requirements must be followed. Article 300 also tells us where a box or fitting is required and where a box only is permitted. *NEC*® 300.15(A) states that all fittings and connectors must be designed for the specific wiring method for which they are used and listed for that purpose. Look up the definition for a branch-circuit listed in Article 100 now. An important requirement for all wiring methods is found in 300.21. The requirement limits the spread of fire or products of combustion and requires that the wiring system be installed so as to minimize the spread of fire and products of combustion; that all penetrations through floors, walls, and ceilings be sealed in an approved manner to accomplish this purpose. *NEC*® 300.22, which contains the provisions for the installation and uses of electrical wiring and equipment in ducts, plenums, and other air-handling spaces, states these sections are safety concerns in determining the wiring method to be used in electrical installations; these should be studied carefully before making

the final selections. Part II of Article 300 explains the requirements for installations over 600 volts nominal. *NEC*® 300.50 covers the underground installations and references, which are the general wiring requirements for installations over 600 volts in the table and section in 300.50. Other general articles are found in Chapter 3 relating to wiring methods. Article 590 covers the temporary wiring requirements needed to provide power for the workers making the installation and constructing the facility during the construction phases of a job. Article 590 also covers other types of temporary wiring, such as temporary wiring concerned with holiday decorative lighting, trade fairs, emergencies and tests, and experimental and developmental work. Be aware of Article 590. In Article 310 you will find all the requirements related to conductors for general wiring. They do not apply to conductors that form an integral part of equipment, such as motors, motor controllers and similar equipment, or conductors specifically provided for elsewhere in the *Code*. Examples would be Article 400 and Article 402 for flexible cords and fixture wires. Other sections dealing with conductors would be Chapter 7 and Chapter 8 of the *NEC*®, which deal with limited voltage wire, such as in

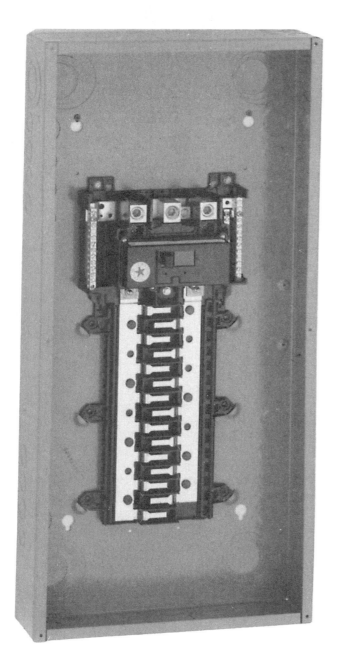

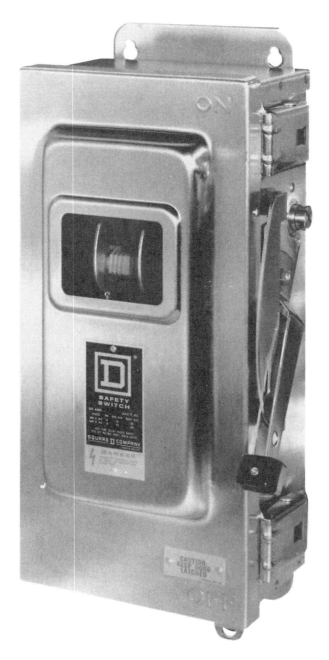

Figure 10–2 Main breaker panelboard interior and enclosure (see 2005 *NEC®* Article 312). *(Courtesy of Square D Company)*

Figure 10–3 Typical safety switch with view safety glass (see 2005 *NEC®* Article 404). *(Courtesy of Square D Company)*

Article 725, 760, and so on. Article 310 must be used in every installation made. The requirements in 310.1 through 310.15 are general requirements for the use and installation of conductors.

Note: When using the tables, read each heading carefully. Read the columns carefully. Read the ambient temperature correction factors below each column, and remember that the notes (footnotes) under each table are a part of those tables and are mandatory requirements. This is important when studying for a test because many test questions will require you to study these additional parts of the table to get the correct answer.

**300.11(A) through (C)
Securing and Supporting**

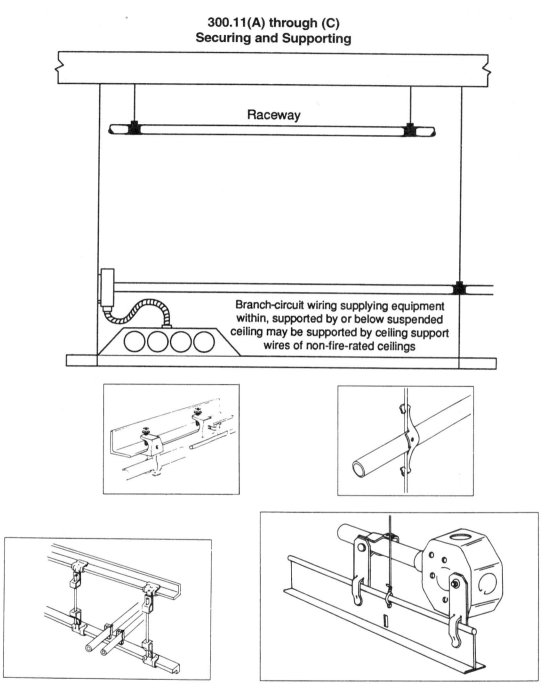

Figure 10–4 Various types of hangers and supporting methods and hardware for raceways. *(Courtesy of Cooper B-Line, Inc.)*

For example, 240.4(D) limits the size overcurrent protective device for small conductors 14 AWG, 12 AWG, and 10 AWG for general wiring not specifically covered by 240.4(A), (B), (C), (E), (F), and (G). These requirements once appeared as obelisk notes under the tables in Article 310. Motors, for instance, generally are not required to comply with 240.4(D) because Article 430 has specific requirements for branch- circuit conductors for motors. Study these tables carefully. The next important part of Article 310 is 310.15(B)(6). Table 310.15(B)(6) has less restrictive ampacity requirements for dwelling services and feeders.

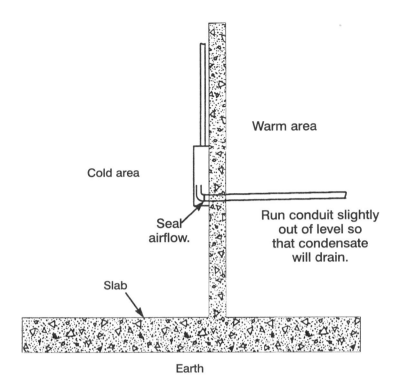

Warm area

Cold area

Seal airflow.

Run conduit slightly out of level so that condensate will drain.

Slab

Earth

Figure 10–5 Seals are required to comply with *NEC*® 300.7 where the raceway is subjected to more than one temperature, such as shown. The system enters the building from outside.

Note: These only apply to single-phase, 120/240 volt, 3-wire systems and do not apply to any other type systems.

Example: NEC® 392.9(A)(1)

Width selection for cable tray containing 600-volt multiconductor cables, sizes 4/0 AWG and larger only. Cable installation is limited to a single layer. The sum of the cable diameters (Sd) must be equal to or less than the usable cable tray width.

Cable tray width is obtained as follows:

A—Width required for 4/0 AWG and larger multi-conductor cables [*see diagram bottom of page 142*]:

Item No.	List Cable Sizes	(D) List Cable Outside Diameter	(N) List Number of Cables	Multiply (D) × (N) Subtotal of Sum of Cable Diameters (Sd)
1	3/C–500 kcmil	2.25 inches	4	9.00 inches
2	3/C–250 kcmil	1.76 inches	3	5.78 inches
3	3/C–4/0 kcmil	1.55 inches	10	15.50 inches

The sum of the diameters (Sd) of all cables (add total Sd for items 1, 2, and 3):

9.00 inches + 5.78 inches + 15.50 inches = 30.28 inches (Sd)

A cable tray with a usable width of 30 inches (750 mm) is required. For approximately 15 percent more, a 36-inch (900 mm) wide cable tray could be purchased that would provide for some future cable additions.

Notes:

1. Cable sizes used in this example are a random selection.
2. Cables—copper conductor with cross-linked polyethylene insulation and a PVC jacket. (These cables could be ordered with or without an equipment grounding conductor.)
3. Total cable weight per foot for this installation:
   61.4 lb/ft (without equipment grounding conductors)
   69.9 lb/ft (with equipment grounding conductors)

This load can be supported by a load symbol "B" cable tray—75 lb/ft.

Table 352.44(A) Expansion Characteristics of PVC Rigid Nonmetallic Conduit Coefficient of Thermal Expansion = 6.084 × 10⁻⁵ mm/mm/°C (3.38 × 10⁻⁵ in./in./°F)

Temperature Change (°C)	Length Change of PVC Conduit (mm/m)	Temperature Change (°F)	Length Change of PVC Conduit (in./100 ft)	Temperature Change (°F)	Length Change of PVC Conduit (in./100 ft)
5	0.30	5	0.20	105	4.26
10	0.61	10	0.41	110	4.46
15	0.91	15	0.61	115	4.66
20	1.22	20	0.81	120	4.87
25	1.52	25	1.01	125	5.07
30	1.83	30	1.22	130	5.27
35	2.13	35	1.42	135	5.48
40	2.43	40	1.62	140	5.68
45	2.74	45	1.83	145	5.88
50	3.04	50	2.03	150	6.08
55	3.35	55	2.23	155	6.29
60	3.65	60	2.43	160	6.49
65	3.95	65	2.64	165	6.69
70	4.26	70	2.84	170	6.90
75	4.56	75	3.04	175	7.10
80	4.87	80	3.24	180	7.30
85	5.17	85	3.45	185	7.50
90	5.48	90	3.65	190	7.71
95	5.78	95	3.85	195	7.91
100	6.08	100	4.06	200	8.11

Table 352.44(B) Expansion Characteristics of Reinforced Thermosetting Resin Conduit (RTRC) Coefficient of Thermal Expansion = 2.7 × 10⁻⁵ mm/mm/°C (1.5 × 10⁻⁵ in./in./°F)

Temperature Change (°C)	Length Change of RTRC Conduit (mm/m)	Temperature Change (°F)	Length Change of RTRC Conduit (in./100 ft)	Temperature Change (°F)	Length Change of RTRC Conduit (in./100 ft)
5	0.14	5	0.09	105	1.89
10	0.27	10	0.18	110	1.98
15	0.41	15	0.27	115	2.07
20	0.54	20	0.36	120	2.16
25	0.68	25	0.45	125	2.25
30	0.81	30	0.54	130	2.34
35	0.95	35	0.63	135	2.43
40	1.08	40	0.72	140	2.52
45	1.22	45	0.81	145	2.61
50	1.35	50	0.90	150	2.70
55	1.49	55	0.99	155	2.79
60	1.62	60	1.08	160	2.88
65	1.76	65	1.17	165	2.97
70	1.89	70	1.26	170	3.06
75	2.03	75	1.35	175	3.15
80	2.16	80	1.44	180	3.24
85	2.30	85	1.53	185	3.33
90	2.43	90	1.62	190	3.42
95	2.57	95	1.71	195	3.51
100	2.70	100	1.80	200	3.60

Figure 10–6 (Reprinted with permission from NFPA 70-2005, the *National Electrical Code*®, Copyright © 2004–, National Fire Protection Association, Quincy, MA 02269. This reprinted material is not the complete and official position of the National Fire Protection Association on the referenced subject, which is represented only by the standard in its entirety.)

Example

380 ft. of conduit is to be installed on the outside of a building exposed to the sun in a single straight run. It is expected that the conduit will vary in temperature from 0°F in the winter to 140°F in the summer (this includes the 30°F for radiant heating from the sun). The installation is to be made at a conduit temperature of 90°F. From the table, a 140°F temperature change will cause a 5.7 in. length change in 100 ft. of conduit. The total change for this example is 5.7″ × 3.8 = 21.67″ which should be rounded to 22″. The number of expansion couplings will be 22 ÷ coupling range (6″ for E945, 2″ for E955). If the E945 coupling is used, the number will be 22 ÷ 6 = 3.67 which should be rounded to 4. The coupling should be placed at 95 ft. intervals (380 ÷ 4). The proper piston setting at the time of installation is calculated as explained above.

$$0 = \left[\frac{140 - 90}{140}\right] 6.0 = 2.1 \text{ in.}$$

Insert the piston into the barrel to the maximum depth. Place a mark on the piston at the end of the barrel. To properly set the piston, pull the piston out of the barrel to correspond to the 2.1 in. calculated above.

See the drawing below.

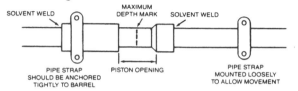

Figure 10–7 Rigid nonmetallic conduit. For aluminum raceway, 21.7 × 0.1 = 2.17 inches an expansion fitting may be required. *(Courtesy of Carlon, a Lamson & Sessions Company)*

Example: NEC® 392.9(A)(2)

Width selection for cable tray containing 600-volt multiconductor cables, sizes 3/0 AWG and smaller. Cable tray allowance fill areas are listed in Column 1 of Table 392.9.
Cable tray width is obtained as follows:

Method 1:
The sum of the total areas for items 1, 2, 3, and 4:

3.34 sq. in. + 3.04 sq. in. + 6.02 sq. in. + 16.00 sq. in. = 28.40 sq. inches

From Table 392.9, Column 1, a 30-inch (750 mm) wide tray with an allowable fill area of 35 square inches must be used. The 30-inch (750 mm) cable tray has the capacity for additional future cables (6.60 square inches additional allowable fill area can be used).

Method 2:
The sum of the total areas for items 1, 2, 3, and 4 multiplied by (6 sq. in./7 sq. in.) = cable tray width required:

3.34 sq. in. + 3.04 sq. in. + 6.02 sq. in. + 16.00 sq. in. = 28.40 sq. inches

$$\frac{28.40 \text{ sq. inches} \times 6 \text{ sq. inches}}{7 \text{ sq. inches}} = 24.34 \text{ inch cable tray width required}$$

Use a 30-inch (750 mm) cable tray.

NEC® 392.9(A)(1)

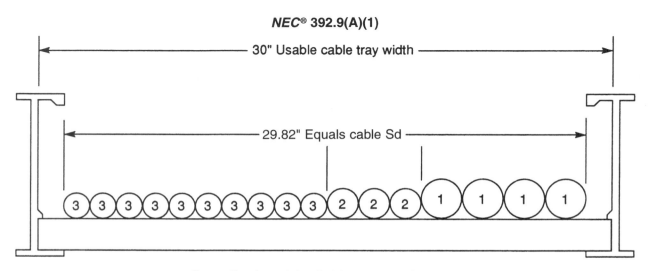

Cross Section of the Cables and the Cable Tray

(Courtesy of Cooper B-Line, Inc.)

NEC® 392.9(A)(2)

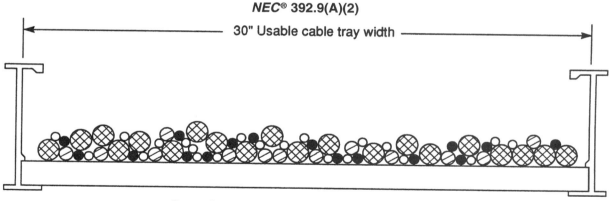

Cross Section of the Cables and the Cable Tray

(Courtesy of Cooper B-Line, Inc.)

Notes:

1. The cable sizes used in this example are a random selection.
2. Cables—copper conductors with cross-linked polyethylene insulation and a PVC jacket. These cables could be ordered with or without an equipment grounding conductor.
3. Total cable weight per foot for this installation: 31.9 lb/ft. (Cables in this example do not contain equipment grounding conductors.) This load can be supported by a load symbol "A" cable tray—50 lb/ft.

Example: NEC® 392.9(A)(3)

Width selection for cable tray containing 600-volt multiconductor cables, sizes 4/0 AWG and larger (single layer required) and 3/0 AWG and smaller. These two groups of cables must have dedicated areas in the cable tray.

Cable tray width is obtained as follows:

A—Width required for 4/0 AWG and larger multiconductor cables:

Item No.	List Cable Sizes	(D) List Cable Outside Diameter	(N) List Number of Cables	Multiply (D) × (N) Subtotal of Sum of Cable Diameters (Sd)
1	3/C–500 kcmil	2.26 inches	3	6.78 inches
2	3/C–4/0 kcmil	1.55 inches	4	6.20 inches

Total cable tray width required for items 1 and 2:

6.78 inches + 6.20 inches = 12.98 inches

B—Width required for 3/0 AWG and smaller multiconductor cables:

Total cable tray width required for items 3, 4, and 5:

(3.20 sq. in. + 4.00 sq. in. + 3.20 sq. in.) (6 sq. in. / 7 sq. in.)1= (10.4 sq. in.) (6 sq. in. / 7 sq. in.)1 = 8.92 inches

Actual cable tray width is A "Width" (12.98 inches) + B "Width" (8.92 inches) = 21.90 inches

A 24-inch (600 mm) wide cable tray is required. The 24-inch (600 mm) cable tray has the capacity for additional future cables (3.1 inches or 3.6 square inches allowable fill can be used).

Notes:

1. This ratio is the inside width of the cable tray in inches divided by its maximum fill area in square inches from Table 392.9, Column 1.
2. The cable sizes used in this example are a random selection.
3. Cables—copper conductor with cross-linked polyethylene insulation and a PVC jacket.
4. Total cable weight per foot for this installation: 40.2 lb/ft (Cables in this example do not contain equipment grounding conductors.) This load can be supported by a load symbol "A" cable tray—50 lb/ft.

Example: NEC® Section 392.9(B)

50 percent of the cable tray usable cross-sectional area can contain type PLTC cables.

4 inches × 6 inches × .500 = 12 square inches allowable fill area

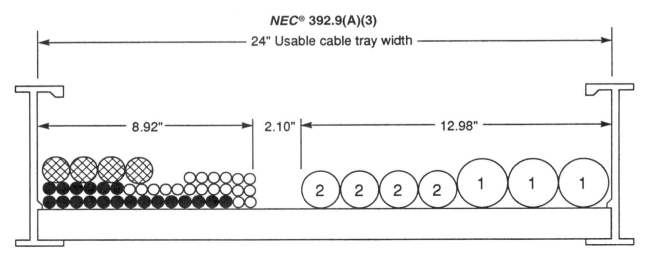

Cross Section of the Cables and the Cable Tray

(Courtesy of Cooper B-Line, Inc.)

2/C 16 AWG 300-volt shielded instrumentation cable O.D. = .224 inch

Cross-Sectional Area = 0.04 square inch

$$\left(\frac{12 \text{ sq. in.}}{0.04 \text{ sq. in. cable}}\right) = \begin{array}{l}300 \text{ cables can be installed} \\ \text{in this cable tray}\end{array}$$

$$\left(\frac{300 \text{ cables}}{26 \text{ cables/row}}\right) = \begin{array}{l}11.54 \text{ rows can be installed} \\ \text{in this cable tray}\end{array}$$

Notes:
1. The cable sizes used in this example are a random selection.
2. Cable—copper conductors with PVC insulation, aluminum/mylar shielding, and PVC jacket.

Following these general application Articles 300 and 310, the remaining articles in Chapter 3 are specific articles. Article 392 covers cable trays (again see Figure 10–4). Cable trays are a support system and not a wiring method.

Note: As you begin to study this part of Chapter 3, refer to the definitions in Article 100 for raceway. A fine print note will alert you to the types of raceways. There is a distinct difference between raceways and cables and other wiring methods. Study that definition or refer to it as needed. A cable tray is neither a raceway nor a cable. It is merely a support system for other wiring methods.

Example: Application of 392.13. Ampacity of Type MV and Type MC Cables (2001 Volts or Over) in Cable Trays—(b) Single Conductor cables. These single conductor cables can be installed in a cable tray cabled together (triplexed, quadruplexed, and so on) if desired. Where the cables are installed according to the requirements of 392.12, the ampacity requirements are as shown in the accompanying chart.

As needed, refer to Articles 320 through 398 for the specific types of raceways and cables. In addition to the references for raceways in Chapter 3, you will find references related to Chapter 9 tables and examples, such as the reference found in 342.22, the numbers of conductors in conduit.

Example: A typical 480/277-volt 400-ampere feeder is run from a distribution unit substation under a concrete slab exposed above a suspended ceiling turning down into a panelboard. Based on the length (200 feet [61 m]) of the run and the calculated load, the installer chooses to pull three 600-kcmil THWN ungrounded circuit conductors and a full-sized 600-kcmil THWN neutral (grounded conductor). If the installer chooses a nonmetallic raceway for a portion of this run, he recognizes that he will also have to add an equipment grounding conductor sized in accordance with Table 250.122 or a 3 AWG THWN copper. However, when he applies the tables in Chapter 9, he finds that he will need the following sizes of conduit or tubing to make this installation:

Sec. No.	Cable Sizes	Solid Unventilated Cable Tray Cover	Applicable Ampacity Tables *	Amp. Table Values By	Special Conditions
(1)	1/0 AWG and Larger	No Cover Allowed **	310.69 and 310.70	0.75	—
(1)	1/0 AWG and Larger	Yes	310.69 and 310.70	0.70	—
(2)	1/0 AWG and Larger in Single Layer	No Cover Allowed **	310.69 and 310.70	1.00	Maintained Spacing of 1 Cable Diameter
(2)	Single Conductor in Triangle Config. 1/0 AWG and Larger	No Cover Allowed **	310.71 and 310.72	1.00	Maintained Spacing of 2.15 × 1 Conductor O.D.

* The ambient ampacity correction factors must be used.
** At a specific position where it is determined that the tray cables require mechanical protection, a single cable tray cover of 6 feet (1.8 m) or less in length can be installed.

NEC® 392.9(B)

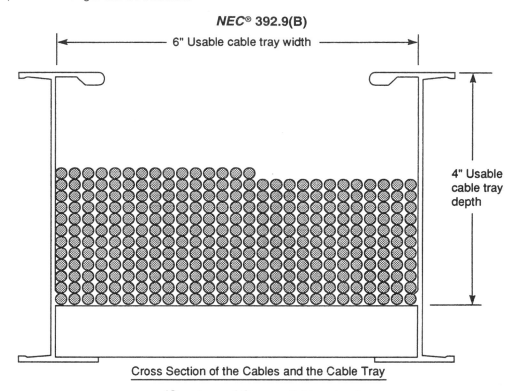

Cross Section of the Cables and the Cable Tray

(Courtesy of Cooper B-Line, Inc.)

Solution:

Table 5. 600-kcmil THWN =
  0.8676 sq. in.   4 × 0.8676   = 3.47 sq. in.
Table 4.
EMT electrical metallic tubing
  40% fill as per Table 1   = 3 (78)
IMC intermediate metal conduit
  40% fill as per Table 1   = 3½ (91)

GRC galvanized rigid conduit
  40% fill as per Table 1   = 3½ (91)

Table 5. 600-kcmil THWN =
  0.8676 sq. in. 4 × 0.8676   = 3.47   sq. in.
Plus an equipment grounding
  conductor 3 AWG THWN   = 0.0973 sq. in.
                          = 3.5673 sq. in.

Table 4.
ENT electrical nonmetallic
tubing max. size 2 (63)   = N/A
Sch. 40 rigid nonmetallic conduit
40% fill as per Table 1   = 3½ (91)
Sch. 80 rigid nonmetallic conduit
40% fill as per Table 1   = 4 (103)

Using these same calculations with 500-kcmil for the ungrounded and grounded conductors would require a trade size 3 (78) conduit for EMT, IMC, GRC, and PVC Schedule 40, and a trade size 3½ (91) for PVC Schedule 80. As one can see, calculations will be necessary to determine the most efficient wiring method.

When sizing the number of conductors in a raceway, conduit, tubing, and so on, it is necessary to refer to the permitted percent fills specified in the tables in Chapter 9, Table 1 for percentage fill and Table 4 for the conduit dimensions. Similar references are found in 344.22 for rigid metal conduit, 352.22 for rigid nonmetallic conduit, 358.22 for electrical metallic conduit, and 348.22 for flexible metallic conduit. Other similar references are found for other raceway types. The number of conductors allowed in a given raceway must be calculated in square inches, based on the insulation and wire size listed in Table 5 and Table 5-A of Chapter 9. For bare conductors, Chapter 8 dimensions are used. Where conductors of all the same size are installed in a single raceway, Table 3 applies generally. The following example will help you learn to apply Chapter 9.

Conduit and tubing diameters will make a difference in the 2002 NEC®.

Trade size 1/2 (16) IMC intermediate metal conduit internal diameter .0060 inch

Allowable 40 percent fill would permit three 8 AWG Type THWN conductors.

Trade size 1/2 (16) Schedule 80 rigid nonmetallic conduit internal diameter 0.526 inch

Allowable 40 percent fill would permit one 8 AWG Type THWN conductor.

Example: What size PVC rigid nonmetallic conduit is required for a multiple branch-circuit supplying a cooling tower located adjacent to a hospital containing three 6 AWG THWN stranded copper conductors, three 4 AWG THWN copper conductors, four 1/0 AWG THW copper conductors, and one 2 AWG bare copper conductor?

Step 1. Rigid nonmetallic conduit, Article 352. Number of conductors, 352.22; Ref. Table 1, Chapter 9.

Step 2. Chapter 9, Notes 1, 2, Table 1 and Notes. Read the notes to the table carefully because they are a part of the table and, where applicable, they are mandatory.

Step 3. Chapter 9 Table 1—Over two conductors permit a 40 percent fill of the raceway.

Step 4. Table 4. Dimensions and percent are of conduit and tubing in square inches. Therefore, select square-inch dimensions from Tables 5 and 8 for the correct calculations.

Step 5.
Table 5.
3 6 AWG THWN = .0507 × 3 = .1521 sq. inch
3 4 AWG THWN = .0824 × 3 = .2472 sq. inch
4 6 AWG THWN = .2223 × 4 = .8892 sq. inch

Table 8.
1 2 AWG BARE = .067 × 1 = .067 sq. inch
                    Total = 1.3555 sq. inches

Table 4.
Not lead covered over two conductors in a raceway 40 percent. Trade size 2 (53) Rigid PVC Schedule 40 raceway is 1.316 square inches. Trade size 2½ (63) raceway is 1.878. Therefore, 2 (53) is not large enough, and it is necessary to install 2½ (63) rigid nonmetallic conduit for this installation.

Note: Trade size 2 (53) (IMC) intermediate metal conduit (Article 342) would permit a 40 percent fill of 1.452 square inches. Therefore, trade size 2 (53) IMC could be used.

Article 314 covers outlet boxes, pull boxes, junction boxes, conduit bodies, and fittings. A conduit body is what is normally referred to in the trade as a Tee, LB, or as a condulet. Use Article 314 carefully, and as you determine the number of conductors for a given box, the sizing requirements will be found in 314.16.

Note: Most boxes with the integral-type clamps, such as nonmetallic outlet boxes, are clamps and a deduction must be added. If uncertain, check with the manufacturer. In a test, normally, the test material will tell you whether or not there are clamps to be calculated. Study 314.16(A) and (B) and the tables carefully.

NEC® 314.16(A)(2)

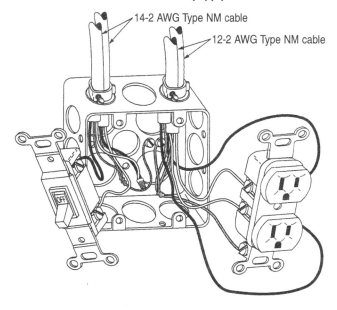

14-2 AWG Type NM cable

12-2 AWG Type NM cable

Example: What size outlet box with integral clamps is required to accommodate a 15-ampere snap switch to control lighting and a 20-ampere duplex receptacle? The switch has a 2-wire w/GRD 14 AWG nonmetallic-sheathed cables feeding it with another 14 AWG NM 2-wire w/ GRD feeding the luminaire (light fixture). The receptacle is fed with a 2-wire w/GRD 12 AWG nonmetallic-sheathed cable and a 2-wire w/GRD 12 AWG nonmetallic sheath cable goes on to feed additional duplex receptacles in the same area.

Solution:

NEC® 314.16(A)(1) and (2). The minimum size of box in cubic inches is based on the following calculations:

The required deduction for each type fitting (i.e., clamps, hickeys, etc.) and all equipment grounding conductors require one deduction for each based on the largest conductor entering the box; 314.16(A) (1), (2), and Table 314.16(B).

In this example, a 12 AWG is the largest, therefore, a 2.25 cubic inch deduction is required for each.

Equipment grounding conductors	=	2.25
Cable clamps = 12 AWG 2.25 × 1	=	2.25

Each device requires two deductions each, based on the size of conductor connected to the device.

Receptacle = 12 AWG	2 × 2.25	=	4.50
Switch = 14 AWG	2 × 2.00	=	4.00

Conductors entering the box, one deduction each.

12 AWG	2.25 × 4	=	9.00
14 AWG	2.00 × 4	=	8.00
			30

Answer: The minimum size box permitted would be a 30 cubic inch box.

In addition, sizing of pull boxes and junction boxes will be included in most examinations. These requirements are found in 314.28.

Example: Four trade size 3 (78) rigid steel conduits or EMT, straight pull. Three trade size 2 (53) rigid steel conduits or EMT, angle pull.

Recommended minimum spacing for 3 (78) conduits, 4¾ inches; for 2 (53) conduits, 3⅜ inches; and between a 3 (78) and a 2 (53), conduit, 4 inches.

For depth of box (A):

Three trade size 2 (53) conduits, angle pull:

A = (6 × 2) + 2 + 2	= 16 inches

Spacing for two rows of conduit:

Edge of box to center of trade size 3 (78) conduit	= 2⅜ inches
Center to center, trade size 3 (78) to trade size 2 (53) conduit	= 4
Minimum distance, D	= 8½
Total	= 14⅞ inches

Required depth: Use 16 inches.

For length of box (B):

Four trade size 3 (78) conduits, straight pull:

B = 8 × 3	= 24 inches

Trade size 2 (53) conduit, angle pull:

B = A	= 16 inches

Required length: Use 24 inches.

For width of box (C):

Recommended spacing for trade size 3 (78) conduits   = 4¾ inches

C = 4 × 4¾ inches	= 19 inches

Required width: Use 19 inches.

Angle pull distance (D):

D = 6 × 2 inches	= 12 inches

Required distance: Use 12 inches.

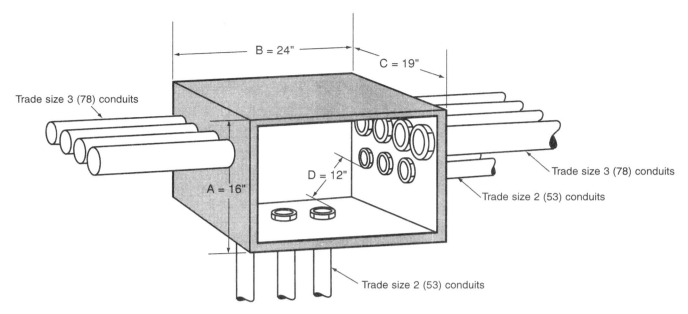

Pull and junction boxes. Minimum size as required in *NEC®* 314.28.

This section should also be studied, and you must know where to find it quickly so that you can determine the size, outlet, or junction box as required.

Article 312 covers cabinet, cutout boxes, and meter socket enclosures. Most panelboards, switches, disconnecting switches, conductors, and so on, are enclosed in a cabinet or cutout box. When determining the requirements for cabinets and cutout boxes, refer to Article 312.

Article 366 covers auxiliary gutters. Auxiliary gutters are not raceways or cables. An auxiliary gutter is an extension used to supplement wiring spaces in meter enclosures, motor control centers, distribution centers, switchboards, and similar wiring systems, and may enclose conductors or bus bars, but is not permitted for switches, overcurrent devices, appliances, or other similar equipment. Although very similar in appearance to a wireway, an auxiliary gutter, has a different application and use. Be aware of Article 366 for auxiliary gutters.

Article 404 covers switches. In this article you can find requirements for all types of switches and some installation requirements, such as accessibility and grouping. Article 408 covers switchboards and panelboards. This article will undoubtedly be included in most examinations and includes many requirements for panelboards and switches. Of particular importance is 110.26(F), an installation requirement. You are urged to mark in your *Code* book 110.26(F) to also refer to 110.26(F) for the dedicating working space above and below equipment. This section also will be in most examinations. You need not try to memorize this section, because it is specific requirements for this equipment, and your proficiency in using this as a reference document will adequately guide you through these wiring method articles. Be familiar with the article names as listed in the Table of Contents, and it will assist you in using this book as a reference or as a tool in the field, and will greatly increase your proficiency in preparing for an examination.

QUESTION REVIEW

> Each lesson is designed purposely to require the student to apply the entire *NEC*® text, and not specific chapters or articles. It has been found that when studying for a timed, open-book examination, the student must gain proficiency in the Table of Contents, the Index, and the ability to move quickly from cover to cover to find the correct answer to each question in a timely fashion.

1. When installing Type SER service-entrance cable in a building from the service panel to a dryer outlet, what are the support requirements for this cable?

   Answer: _____

   _____

   _____

   _____

   Reference: _____

2. When installing conduit bodies that are legibly marked with their cubic inch capacity, how can one determine the number of conductors and splices and taps permitted?

   Answer: _____

   _____

   _____

   _____

   Reference: _____

3. Is trade size ⅜ (12) liquidtight flexible metal conduit permitted for enclosing motor leads in accordance with 430.245(B)?

   Answer: _____

   _____

   _____

   _____

   Reference: _____

4. Structural bar joists are often spaced at up to 5-foot (1.5 m) intervals. Support is difficult to achieve when installing metal raceways on these bar joists. Is it permitted to mount an outlet box on a bar joist and extend intermediate metal conduit (IMC) 5 feet (1.5 m) to the first support?

   Answer: _____

   _____

   _____

   _____

   Reference: _____

5. When installing electrical nonmetallic tubing (ENT), is it permissible to install this material through metal studs without securing the ENT every 3 feet (900 mm) as required by the *Code*?

   Answer: _____

   _____

   _____

   _____

   Reference: _____

6. What are the support requirements and methods of securing Type NM cable?

   Answer: _____

   _____

   _____

   _____

   Reference: _____

7. Can nonmetallic wireways be installed in hazardous Class 1 or Class 2 locations?

   Answer: _____

   _____

   _____

   _____

   Reference: _____

8. What wiring methods are permitted for installing a branch-circuit routed through the bar joists of a metal building for a 4160-volt, 3-phase motor?

   Answer: _____

   _____

   _____

   _____

   Reference: _____

9. A contract to install a branch-circuit under a bank president's desk to serve a calculator is awarded to a contractor. The bank has asked that there be no drilled holes in the walls. Can the conductors be laid on the floor under the carpet? If so, what type of wiring method should be used?

   Answer: _____

   _____

   _____

   _____

   Reference: _____

Chapter Ten • 151

10. When installing wireway metallic or nonmetallic, either type, how do you derate the conductors when exceeding the number permitted?

    Answer: _____

    _____

    _____

    _____

    Reference: _____

11. When installing Type AC cable on wood studs, routing them parallel to the framing members, what distance clearance must be maintained where the Type AC cable is likely to be penetrated by screws or nails?

    Answer: _____

    _____

    _____

    _____

    Reference: _____

12. How do the installation requirements for intermediate metal conduit (IMC) and that of rigid metal conduit (RMC) differ? What installation requirement is permitted by one wiring method that is not permitted by the other?

    Answer: _____

    _____

    _____

    _____

    Reference: _____

13. What is the minimum size conductor that can be run in parallel, generally? (Do not consider exceptions.)

    Answer: _____

    _____

    _____

    _____

    Reference: _____

14. Where installing electrical nonmetallic tubing for the branch-circuit conductors in a five-story building, does the material from which the walls, floors, and ceilings are constructed need to be considered?

    Answer: _____

    _____

    _____

    _____

    Reference: _____

15. Christmas tree lights have been installed on the local courthouse. The mayor suggested that we leave them up throughout the year. Is this permitted? How long are they permitted to remain installed?

Answer: _____

_____

_____

_____

Reference: _____

16. When installing a run of conduit, it is necessary to install a piece of liquidtight flexible metal conduit 6 feet (1.8 m) long through a plenum. Is this permitted by the *Code*? How many feet of liquidtight flexible metal conduit can you install in a plenum?

Answer: _____

_____

_____

_____

Reference: _____

17. Where making an installation of rigid nonmetallic conduit, Schedule 80, along the outside of a building to feed outdoor floodlights and studying the *Code*, you find that in 300.7(B) expansion joints must be provided where necessary. How much expansion or contraction must rigid nonmetallic conduit have before an expansion fitting is required?

Answer: _____

_____

_____

_____

Reference: _____

18. A column panel is a narrow panel that is installed inside or flush with an I-beam. It is primarily used in warehouses and areas where the panel is subjected to moving traffic and warehousing storage, such as forklifts and palletizing. When installing a column-type panel, an auxiliary gutter is installed above the panel, and the neutral connections are made up at the top of the auxiliary gutter and not brought down into the panelboard. Is this an acceptable wiring method?

Answer: _____

_____

_____

_____

Reference: _____

19. When wiring a single-family dwelling to save space, a trade size 2 (53) rigid conduit is installed from the top of the panelboard into the attic space and nonmetallic-sheathed cables are pulled within that trade size 2 (53) conduit. Is this a permitted wiring method?

    Answer: _____

    _____

    _____

    _____

    Reference: _____

20. A small lake cabin is to be wired. This cabin will contain wood heating, no air conditioning, and only six single-pole branch-circuits for lighting and receptacle outlets. Using only six switches in this panel, are you therefore exempt from using a main circuit breaker or a single disconnecting switch as required in Chapter 2 for services?

    Answer: _____

    _____

    _____

    _____

    Reference: _____

Chapter Eleven

OBJECTIVES

Studying this chapter along with the 2005 *NEC®* gives the reader a basic introduction to the branch-circuit and feeder requirements in the *Code* that apply to all installations for above and below 600 volts.

After studying this chapter, you should know:

- The definitions and the application of terms that are applicable and that are unique to the 2002 *NEC®*

- The new *Code* articles that have been relocated to Chapter 4 and are addressed as they apply to the *NEC®*

- The utilization equipment requirements found in this chapter

- The flexible cords and fixture wire requirements and applications

- The switch types, requirements, and applications

- The new Article 406 covering all requirements for receptacles, cord connectors, and attachment plugs (caps)

- The switchboard and panelboard requirements and applications

- The basic requirements for luminaires (lighting fixtures), motors, transformers, and other general-use equipment commonly found in all installations

GENERAL-USE UTILIZATION EQUIPMENT

The equipment covered in Chapter 4 is generally found in most facilities—residential, commercial, and industrial. Chapter 4 begins with two articles about wire. One might think that these articles are misplaced and should be found in Chapter 3. However, flexible cords, cables, and fixture wires are usually associated with general-purpose equipment, such as flexible cords for appliances and motors, and fixture wiring is usually associated with luminaires (lighting fixtures).

1. The maximum wattage permitted for a medium base incandescent lamp (standard light bulb) is _____ watts.
2. The maximum wattage permitted for a mogul-base incandescent lamp is _____ watts.
3. A special base or other means must be used for incandescent lamps of _____ watts and larger.

Answers: 1. 300   2. 1500   3. 1501
Reference: *NEC®* 410.53

Chapter 4 covers all types of luminaires (lighting fixtures), appliances, fixed electric space heating equipment, motors, transformers, air-conditioning and refrigeration equipment, and several special use types of general equipment, such as phase converters often used in rural farm-type installations, and capacitors now generally found in many types of equipment that improve the efficiency of many services as a power factor correction. Generators are often used to supplement or provide emergency power or standby power. (See Article 600 for electric signs and outline lighting.) Article 422, which covers the appliances, generally has specific requirements for branch-circuits, installation of appliances, and the control and protection of appliances. Space heating, motors, air-conditioning and refrigeration equipment articles also have general provisions and installation provisions within those articles.

NEC® 422.12 states that central heating equipment must be supplied by an "**individual branch circuit**." What is an individual branch circuit?

Answer: An individual branch circuit is defined in Article 100 as a branch circuit that supplies only one utilization equipment.

In the 2005 *NEC®*, Chapter 4 has a new look with new articles. Article 380—Switches is now Article 404 in the 2005 *NEC®*. This revision is a sensible change because switches are electrical equipment and will be

Figure 11–1 Examples of common receptacles, cord connectors, and attachment plugs (caps) covered in Article 406.

easier to locate for anyone new to the *Code*. Article 408 covers switchboards and panelboards. Article 406—Receptacles, Cord Connectors, and Attachment Plugs (Caps) was added to the 2002 *NEC*®. This change was made in an attempt to make the *Code* more user friendly. (See Figure 11-1, Figure 11-2, and Figure 11-3.) Article 410 now only covers luminaires (lighting fixtures), lampholders, and lamps. However, Article 410 does not cover signs.

Would it be permissible to connect a waste (garbage) disposal and a built-in dishwasher with flexible cords? Both cords would be a minimum of 42 inches (1.1 m).

Answer: NEC® 422.16(B)(2) would permit the built-in dishwasher to be connected with a cord 3 feet (900 mm) to 4 feet (1.2 m) long. 422.16(B)(1), however, would prohibit the 42-inch (1.1 m) cord on the waste disposal. This section limits the cord for disposals to 18 to 36 inches (450 to 900 mm).

Note: Portable dishwashers are supplied by the manufacturer with a cord and plug attached. This cord is evaluated as a part of the listing, and length is not limited by 422.16.

Figure 11–2 Compliance with 406.8. *(Courtesy of TayMac Corporation)*

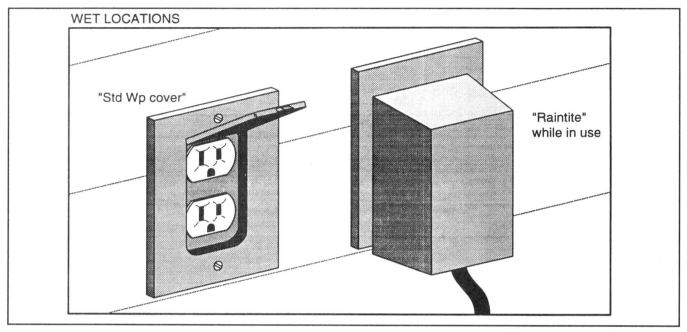

WET LOCATIONS

"Std Wp cover"

"Raintite" while in use

Figure 11–3 Weatherproof cover on the left meets 406.8. The weatherproof cover on the right meets the general rule 406.8. *(Courtesy of TayMac Corporation)*

Article 490 contains requirements for equipment over 600 volts. It covers the general requirements for equipment operating at more than 600 volts and defines "high voltage" as more than 600 volts.

Read *NEC®* 430.6 Carefully

The motor nameplate current is *not* to be used. The values in Tables 247 through 250 are to be used instead.

Example: Motor branch-circuit conductors ampacities are as per Article 310. However, 240.4(D) limiting 14 AWG to a 15-ampere overcurrent device, 12 AWG to a 20-ampere device, and 10 AWG to a 30-ampere device does *not* apply. 430.6 only references the table and not 240.4(D).

Example: In a steel mill, a number of small 3-horsepower, 3-phase, 480-volt motors are supplied by a 400-ampere feeder. What is the minimum size 10-foot (3.0 m) tap conductor permitted to be used to feed each motor?

Step 1. *NEC®* 240.3(E), which permits tap conductors to be protected against overcurrent in accordance with 210.19(A)(3), 240.5 (B)(2), 240.21, 368.17(B), 368.17(C), and 430.53(D).

Step 2. *NEC®* 240.21(F) references 430.28 and 430.53(D) for motor taps.

Step 3. *NEC®* 430.28, 400 amperes ÷ 1000% = 40 amperes

Step 4. Table 310.16

Answer: 8 AWG THWN

Note: Some would permit a 10 AWG conductor to be used in this calculation as taken from the 90° column because a 3-horsepower motor (Table 430.150) would only draw 4.8 amperes, and the temperature would not be exceeded.

QUESTION REVIEW

> **Each lesson is designed purposely to require the student to apply the entire *NEC*® text, and not specific chapters or articles. It has been found that when studying for a timed, open-book examination, the student must gain proficiency in the Table of Contents, the Index, and the ability to move quickly from cover to cover to find the correct answer to each question in a timely fashion.**

THREE-PHASE MOTOR AS FOLLOWS:

Type:	Squirrel cage, high reactance
Voltage:	460 volts, 3-phase
HP:	30
Current:	Nameplate is the same as *NEC*® current Tables 430.247–250
Duty:	Continuous
Start:	High reactance
Code letter:	None
Service factor:	None

Refer to the description of the 3-phase motor above for Questions 1–3.

1. What *NEC*® section requires that the Tables 430.247–250 be used instead of actual current rating marked on the motor nameplate?

   Answer: _____

   _____

   _____

   _____

   Reference: _____

2. What size inverse-time circuit breaker is needed to provide the maximum branch-circuit, short-circuit, ground-fault protection?

   Answer: _____

   _____

   _____

   _____

   Reference: _____

3. A dual element (time delay) fuse can be sized a maximum of _____ percent when used as branch-circuit protection.

   Answer: _____

   _____

   _____

   _____

   Reference: _____

4. Where branch-circuit conductors are tapped to 12 AWG with a 20-ampere circuit rating as permitted in the *NEC®*, the tap cannot be smaller than size _____.

Answer: _____

_____

_____

Reference: _____

In Questions 5 through 8, calculate the ampere rating of the fuse size needed for protection of the following copper conductors where no motors are connected:

5. 12 AWG THHN _____ ampere rated fuse

Answer: _____

_____

_____

Reference: _____

6. 8 AWG THW _____ ampere rated fuse

Answer: _____

_____

_____

Reference: _____

7. 3 AWG THW _____ ampere rated fuse

Answer: _____

_____

_____

Reference: _____

8. 3/0 AWG THWN _____ ampere rated fuse

Answer: _____

_____

_____

Reference: _____

9. For a transformer 600 volts or less that has no secondary overcurrent protection, the primary overcurrent protection shall not exceed _____ percent of the primary current.

    Answer: _____

    _____

    _____

    _____

    Reference: _____

10. The maximum branch-circuit and ground-fault protection using time-delay fuses for typical motors are sized at _____ percent of the motor's full-load current.

    Answer: _____

    _____

    _____

    _____

    Reference: _____

11. Overload protection (heaters, relays, time delay fuses, thermal overloads) for typical motors having a service factor of 1.15 is generally sized at _____ percent of the motor's full-load current.

    Answer: _____

    _____

    _____

    _____

    Reference: _____

12. A motor controller that is installed with the expectation of its being submerged in water occasionally for short periods shall be installed in a rated enclosure type _____ AWG.

    Answer: _____

    _____

    _____

    _____

    Reference: _____

13. Where recessed high-intensity discharge fixtures are installed _____ and operated by remote ballasts, both the fixture and the ballast require thermal protection.

    Answer: _____

    _____

    _____

    _____

    Reference: _____

14. Any pipe or duct system foreign to the electrical installation must not enter a transformer vault. The _____ piping is not considered foreign to the vault.

Answer: _____

_____

_____

_____

Reference: _____

15. A 10 kVA dry-type transformer rated at 480 volts can be installed on a building column and shall not be required to be _____.

Answer: _____

_____

_____

_____

Reference: _____

16. A capacitor is located indoors. It must be enclosed in a vault if it contains more than a minimum of _____ gallon(s) of flammable liquid.

Answer: _____

_____

_____

_____

Reference: _____

17. A single-phase hermetic refrigerant motor-compressor has a rated load current of 24 amperes and a branch-circuit selection current of 30 amperes. The branch-circuit conductors are copper, Type TW. They operate at 80°F, and they are the only conductors in the conduit to this compressor. The smallest possible branch-circuit conductors must be at least size _____ AWG.

Answer: _____

_____

_____

_____

Reference: _____

18. A dry-type transformer is to be installed indoors. If rated more than _____ kilovolt amperes, the transformer must be installed in a fire-resistant transformer room.

Answer: _____

_____

_____

_____

Reference: _____

19. An electric resistance heater is rated for 2400 watts at 240 volts. What power is consumed when the heater is operated at 120 volts?

    Answer: _____

    _____

    _____

    _____

    Reference: _____

20. A 230-volt, single-phase circuit has 10 kilowatts of load and 50 amperes of current. The power factor is _____ percent.

    Answer: _____

    _____

    _____

    _____

    Reference: _____

Chapter Twelve

SPECIAL EQUIPMENT AND OCCUPANCIES

As you begin to study Chapter 12, remember that Chapter 11 covered general equipment, equipment that is found in most every common installation, equipment that is for general use, such as luminaires (light fixtures), receptacles, motors, transformers, and so forth. In this chapter, we note that we are studying special occupancies, special equipment, and special conditions. These rules apply to these items and are not generally found in all buildings. They are special. Study carefully Article 90, Chapter 1, and Article 300 of the text. *NEC®* 90.3 outlines the use of the *NEC®* and states that Chapters 1 through 4 apply generally and Chapters 5, 6, and 7 amend, supplement, or modify Chapters 1 through 4.

> **Caution:** *NEC®* 90.3, Chapter 5 only modifies and augments Chapters 1 through 4. Corrosion, Capacity, Suitability, and Service are not considered. Therefore, it is the responsibility of the designer to take these factors into account along with the hazardous location requirements covered in Chapter 5.

Special occupancies include such things as hazardous areas, bulk storage plants, service stations, paint spray booths, commercial garages, and many more hazardous occupancies. They also include occupancies such as hospitals, mobile homes, and agriculture buildings. On the other hand, Chapter 6 deals with special equipment, such as signs, not found in every building but still common equipment. Swimming pools, welding equipment, electric welders, elevators, dumbwaiters, office furnishings, and many more types are covered.

Chapter 7 deals with special conditions. These are conditions that are special and not general conditions, such as emergency systems, and legally and optional standby systems, over-600-volt systems, optical fiber cables, and raceways. Finally, in Chapter 7 is Article 780, which deals with the intelligent buildings, such as Smart House, the closed loop systems that offer a totally different concept to the safe use of electricity as we have known it in the past. Study the questions carefully. Remember as you study these chapters, they modify, supplement, or amend Chapters 1 through 4 of the *NEC®*.

- Article 600—Signs and Outline Lighting
- Article 605—Office Furnishings
- Article 610—Cranes and Hoists
- Article 620—Elevators, Dumbwaiters, Escalators, Moving Walks, Wheelchair Lifts, and Stairway Chairlifts
- Article 625–Electric Vehicle Charging System Equipment. This article places installation requirements for electrically powered automobiles in the *Code*. This article does not cover other charging systems, such as for golf carts, forklifts, etc. It does cover individual residential-type chargers and the proposed commercial quick-charging stations that may some day appear at rest stops along the interstate highway system. It is said that someday soon you will be able to stop for coffee and get your electric car charged in a few minutes.
- Article 630—Electric Welders
- Article 640—Sound Recording and Similar Equipment
- Article 645—Information Technology Equipment (formerly Computer/Data Processing Equipment)
- Article 647—Sensitive Electronic Equipment
- Article 650—Pipe Organs
- Article 660—X-ray Equipment
- Article 665 Induction and Dielectric Heating Equipment
- Article 668—Electrolytic Cells
- Article 669—Electroplating
- Article 670—Industrial Machinery
- Article 675—Electrical-Driven or Controlled Irrigation Machinery
- Article 680—Swimming Pools, Fountains, and Similar Equipment
- Article 685—Integrated Electrical Systems
- Article 690—Solar Photovoltaic Systems
- Article 692—Fuel Cell Systems
- Article 695—Fire Pumps. This article has added installation requirements into the *Code* that were formerly found in NFPA 20.

QUESTION REVIEW

Each lesson is designed purposely to require the student to apply the entire *NEC*® text, and not specific chapters or articles. It has been found that when studying for a timed, open-book examination, the student must gain proficiency in the Table of Contents, the Index, and the ability to move quickly from cover to cover to find the correct answer to each question in a timely fashion.

1. Name three types of optical fiber cable.

   Answer: _____

   _____

   _____

   _____

   Reference: _____

2. Fixed wiring over a Class I location in a commercial garage cannot be enclosed in nonmetallic-sheathed cables. True or False?

   Answer: _____

   _____

   _____

   _____

   Reference: _____

3. Are seals required for an outdoor propane-dispensing unit located 50 feet (15.24 m) from the office where the branch-circuit supplying the unit originates? Where are the seals required to be located?

Answer: _____

_____

_____

_____

Reference: _____

4. Emergency electrical systems are those systems legally required and classed as emergency by Article 700 *NEC*. True or False?

Answer: _____

_____

_____

_____

Reference: _____

5. In a commercial garage, Class _____, Division _____ wiring methods must be met for an electrical outlet installed 12 inches (300 mm) above the floor if there is no mechanical ventilation.

Answer: _____

_____

_____

_____

Reference: _____

6. According to the *NEC*, transfer equipment for emergency systems shall be designed and installed to prevent _____ of normal and emergency sources of power.

Answer: _____

_____

_____

_____

Reference: _____

7. The sealing compound in a completed conduit seal for a Class I location must be at least _____ inch(es) (_____ mm) thick.

Answer: _____

_____

_____

_____

Reference: _____

8. Bare conductors are field-connected to fixed terminals of different phases on an outdoor, 13.8-kV circuit. According to the *National Electrical Code*®, the air separation between these conductors shall be a minimum of _____ inch(es) (_____ mm).

   Answer: _____

   _____

   _____

   _____

   Reference: _____

9. In a hospital's general care areas, the number of receptacles required to be in a patient-bed location is a minimum of _____.

   Answer: _____

   _____

   _____

   _____

   Reference: _____

10. A 120-volt, single-phase store sign panel supplies only the sign circuits. It is on 24 hours a day. It supplies three ⅓ horsepower motors, 36-0.80 ampere ballasts. Each ungrounded conductor in the subfeeder to the sign panel must be at least _____ AWG copper THWN.

    Answer: _____

    _____

    _____

    _____

    Reference: _____

11. Portable gas tube signs for interior use can be connected with a supply cord having a maximum length of _____ feet (_____ m).

    Answer: _____

    _____

    _____

    _____

    Reference: _____

Answer Questions 12 through 15 about Type S fuses True or False.

12. A 15-ampere Type S fuse will fit into a 20-ampere Type S adapter.

    Answer: _____

    _____

    _____

    _____

    Reference: _____

13. A 20-ampere Type S fuse will fit into a 20-ampere Type S adapter.

Answer: _____

_____

_____

_____

Reference: _____

14. A 25-ampere Type S fuse will fit into a 30-ampere Type S adapter.

Answer: _____

_____

_____

_____

Reference: _____

15. A 6¼-ampere Type S adapter will accept a 3-ampere, 4-, 4½-, 5-, 5⁶⁄₁₀-, and 6¼-ampere Type S fuse.

Answer: _____

_____

_____

_____

Reference: _____

16. The *NEC®* requires that the available fault current be marked when series rated systems are installed. (a) Who is required to install that marking? (b) What must the marking state? (c) Who is responsible for providing this information?

Answer: _____

_____

_____

_____

Reference: _____

17. A 1½-horsepower, single-phase motor that has an efficiency of 80 percent, operates at 230 volts and has an input current of _____ amperes. (1 hp = 746 watts)

Answer: _____

_____

_____

_____

Reference: _____

18. A 30-horsepower, wound-rotor induction motor with no code letter is to be installed with 460-volt, 3-phase, alternating current. Disregarding all the exceptions, the nontime delay fuse for short-circuit protection of the motor branch-circuit must be rated at a maximum _____ amperes.

    Answer: _____

    _____

    _____

    _____

    Reference: _____

19. The following 480-volt, 3-phase, 3-wire, intermittent-use equipment is in a commercial kitchen: two 5000-watt water heaters, four 3000-watt fryers, and two 6000-watt ovens. Each ungrounded conductor in the feeder circuit for this kitchen equipment must be sized to carry a minimum calculated load of _____ amperes.

    Answer: _____

    _____

    _____

    _____

    Reference: _____

20. A 240/480-volt, 3-phase power panelboard supplies only one 15,000 VA, 480-volt, 3-phase balanced resistive load. Each ungrounded conductor in the subfeeder to this power panel has a total net calculated load of _____ amperes.

    Answer: _____

    _____

    _____

    _____

    Reference: _____

Chapter Thirteen

SPECIAL CONDITIONS AND COMMUNICATION SYSTEMS

Special Conditions

The first three articles in Chapter 7 cover Emergency Systems (Article 700), Legally Required Standby Systems (Article 701), and Optional Standby Systems and Other Special Systems (Article 702). The determination for which type system is installed in a building or facility is made by others—fire marshal, building official, or the like. The *NEC*® does not make these requirements; it does not require exit lights in buildings. It does not require smoke detectors. How-

Class 1 Circuit. The portion of the wiring system between the load side of the overcurrent device or power-limited supply and the connected equipment. The voltage and power limitations of the source are in accordance with 725.21.

Class 2 Circuit. The portion of the wiring system between the load side of a Class 2 power source and the connected equipment. Due to its power limitations, a Class 2 circuit considers safety from a fire initiation standpoint and provides acceptable protection from electric shock.

Class 3 Circuit. The portion of the wiring system between the load side of a Class 3 power source and the connected equipment. Due to its power limitations, a Class 3 circuit considers safety from a fire initiation standpoint. Since higher levels of voltage and current than for Class 2 are permitted, additional safeguards are specified to provide protection from an electric shock hazard that could be encountered.

> *725.41(A) FPN No. 2: Table 11(A) and Table 11(B) in Chapter 9 provide the requirements for listed Class 2 and Class 3 power sources.*

III. Class 2 and Class 3 Circuits

725.41 Power Sources for Class 2 and Class 3 Circuits.

(A) Power Source. The power source for a Class 2 or a Class 3 circuit shall be as specified in 725.41(A)(1), (A)(2), (A)(3), (A)(4), or (A)(5)

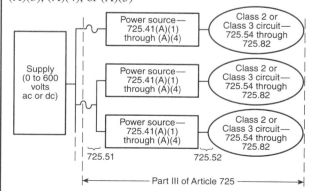

Figure 725.41 Class 2 and Class 3 Circuits.

Reprinted with permission from NFPA 70-2005. *(continued)*

The power source types for a Class 2 or a Class 3 circuit shall be as specified below:

725.41(A)
(1) A listed Class 2 or Class 3 transformer
(2) A listed Class 2 or Class 3 power supply
(3) Other listed equipment marked to identify the Class 2 or Class 3 power source

Exception No. 1: Thermocouples shall not require listing as a Class 2 power source.
Exception No. 2: Limited power circuits of listed equipment where these circuits have energy levels rated at or below the limits established in Chapter 9, Table 11(A) and Table 11(B).

> *FPN: Examples of other listed equipment are as follows:*
> *(1) A circuit card listed for use as a Class 2 or Class 3 power source where used as part of a listed assembly*
> *(2) A current-limiting impedance, listed for the purpose, or part of a listed product, used in conjunction with a non–power-limited transformer or a stored energy source, for example, storage battery, to limit the output current*
> *(3) A thermocouple*
> *(4) Limited voltage/current or limited impedance secondary communications circuits of listed industrial control equipment*

(4) Listed information technology (computer) equipment limited power circuits.

> *FPN: One way to determine applicable requirements for listing of information technology (computer) equipment is to refer to UL 1950–1995, Standard for Safety of Information Technology Equipment, Including Electrical Business Equipment. Typically such circuits are used to interconnect information technology equipment for the purpose of exchanging information (data).*

(5) A dry cell battery shall be considered an inherently limited Class 2 power source, provided the voltage is 30 volts or less and the capacity is equal to or less than that available from series connected No. 6 carbon zinc cells.

725.41(B) Interconnection of Power Sources. *Class 2 or Class 3 power sources shall not have the output connections paralleled or otherwise interconnected unless listed for such interconnection.*

Reprinted with permission from NFPA 70-2005.

ever, when these items are required to be installed, the *NEC*® governs. The installation requirements must conform to these articles as appropriate.

Article 720 covers the *few* circuits and equipment operating at less than 50 volts that are not already covered in the *Code* by Articles 411, 551, 650, 669, 690, 725, and 760.

Article 725 covers power-limited circuits of Class 1, 2, and 3. It is very important that one understand the parameters of each class of wiring system. The power is limited by the power source, generally a transformer.

Article 780—Closed-Loop and Programmed Power Distribution, often referred to as "Smart House," applies to premise power distribution systems jointly controlled by a signaling between the energy-controlling equipment and utilization equipment. All equipment, including the conductors, is required to be listed and identified. It will apply only for signaling circuits not exceeding 24 volts; the current required shall not exceed 1 ampere where protected by an overcurrent device or an inherently limited power source. The article also contains specific safety requirements for this new system.

Listed hybrid cable is permitted that may contain power, communications, and signaling conductors under a common jacket. The jacket must be applied so as to separate the power conductors from the communications and signaling conductors. The individual conductors of a hybrid cable shall conform to the *Code* provisions applicable to their current, voltage, and insulation rating. The signaling conductors shall not be smaller than 24 AWG copper. The power, communications, and signaling conductors of listed hybrid cable are permitted to occupy the same cabinet, panel, or outlet box (or similar enclosure housing the electrical terminations of electric light or power circuits) only if connectors specifically listed for hybrid cable are employed.

770.6 Raceways for Optical Fiber Cables. *Raceways for optical fiber cable shall comply with 770.6(A) through (D).*
(A) Listed Chapter 3 Raceways. *Listed optical fiber cable shall be permitted to be installed in any type of listed raceway permitted in Chapter 3 where that listed raceway is installed in accordance with Chapter 3. Where optical fiber cables are installed within raceway without current-carrying conductors, the raceway fill tables of Chapter 3 and Chapter 9 shall not apply. Where nonconductive optical fiber cables are installed with electric conductors in a raceway, the raceway fill tables of Chapter 3 and Chapter 9 shall apply.*

(continued)

(B) Listed Plenum Optical Fiber Raceway. *Listed optical fiber cable shall be permitted to be installed in listed plenum optical fiber raceway, listed riser optical fiber raceway, or listed general-purpose optical fiber raceway installed in accordance with 770.53 and 362.24 through 362.56, where the requirements applicable to electrical nonmetallic tubing shall apply.*

(C) Listed Riser Optical Fiber Raceway. *Listed plenum optical fiber raceway, listed riser optical fiber raceway, or listed general-purpose optical fiber raceway installed in accordance with 770.53 shall be permitted to be installed as innerduct in any type of listed raceway permitted in Chapter 3.*

(D) Unlisted Innerduct. *Unlisted underground or outside plant construction plastic innerduct entering the building from the outside shall be terminated and firestopped at the point of entrance.*

Reprinted with permission from NFPA 70-2005.

Chapter 8 is a stand-alone chapter in that 90.3 states:

Chapter 8 covers communications systems not subject to the requirements of Chapter 1 through 7 except where the requirements are specifically referenced in Chapter 8.

This chapter is divided into four articles.

Article 800 *covers telephone, telegraph (except radio), outside wiring for fire alarm and burglar alarm, and similar central station systems; and telephone systems not connected to a central station system but using similar types of equipment, methods of installation, and maintenance.*

FPN No. 1: For further information for fire alarm, sprinkler waterflow, and sprinkler supervisory systems, see Article 760.

FPN No. 2: For installation requirements of optical fiber cables, see Article 770.

FPN No. 3: For installation requirements for network-powered broadband communications circuits, see Article 830.

Reprinted with permission from NFPA 70-2005.

See *NEC®* 810.1 for radio and television receiving equipment, amateur radio transmitting and receiving equipment, . . . but not for equipment and antennas used for coupling carrier current to power line conductors.

See *NEC®* 820.1 for coaxial cable distribution of radio frequency signals typically employed in community antenna television (CATV) systems.

Article 830 *covers network-powered broadband communications systems that provide any combination of voice, audio, video, data, and interactive services through a network interface unit.*

FPN No. 1: A typical basic system configuration includes a cable supplying power and broadband signal to a network interface unit that converts the broadband signal to the component signals. Typical cables are coaxial cable with both broadband signal and power on the center conductor, composite metallic cable with a coaxial member for the broadband signal and a twisted pair for power, and composite optical fiber cable with a pair of conductors for power. Larger systems may also include network components such as amplifiers that require network power.

FPN No. 2: See 90.2(B)(4) for installations of broadband communications systems that are not covered.

Reprinted with permission from NFPA 70-2005.

Article 830 covers network-powered broadband communications systems. This is a new generation communication system and provides multimedia services in any combination of audio, voice, video, data, and interactive services. This new system may operate at higher levels of power and recognizes two levels of power.

Network Power Source	Low	Medium
Maximum power rating (volt-amperes)	100	100
Maximum voltage rating (volts)	100	150

Chapter 8 does not reference Table 300.5 (wiring method burial requirements). Therefore, the table cannot be applied. There are no burial depth requirements for communication cables covered in Articles 800, 810, and 820. The reason for that is that the *NEC®* is primarily a safety document. Communications generally pose no safety hazards by interruption of service. Therefore, the installers often bury these cables for service and economic reasons. However, they pose no safety hazards, so the *NEC®* does not set any requirements. Local requirements or amendments may apply.

NEC® 800.8, 820.8, and 830.8 cover the mechanical execution of work for communications circuits, and equipment shall be installed in a neat and professional manner. Cables shall be supported by the building structure in such a manner that the cable will not be damaged by normal building use.

Chapter 8 is unique. Study it carefully. Study the scope as shown throughout this chapter, of each article carefully, to be sure that you are answering a question from the right article and section of this chapter.

NEC® 830.47(C) *Mechanical Protection.* Direct-buried cable, conduit, or other raceways shall be installed to meet the minimum cover requirements of Table 830.47. In addition, direct-buried cables emerging from the ground shall be protected by enclosures, raceways, or other approved means extending from the minimum cover distance re-

---

quired by Table 830.47 below grade to a point at least 2.5 m (8 ft) above finished grade. In no case shall the protection be required to exceed 450 mm (18 in.) below finished grade. Type BMU and BLU direct-buried cables emerging from the ground shall be installed in rigid metal conduit, intermediate metal conduit, rigid nonmetallic conduit, or other approved means extending from the minimum cover distance required by Table 830.47 below grade to the point of entrance.

Exception: A low-power network-powered broadband communications circuit that is equipped with a listed fault protection device, appropriate to the network-powered broadband communications cable used, and located on the network side of the network-powered broadband communications cable being protected.

Reprinted with permission from NFPA 70-2005.

QUESTION REVIEW

Each lesson is designed purposely to require the student to apply the entire *NEC®* text, and not specific chapters or articles. It has been found that when studying for a timed, open-book examination, the student must gain proficiency in the Table of Contents, the Index, and the ability to move quickly from cover to cover to find the correct answer to each question in a timely fashion.

---

1. Does the *Code* address grounding CATV service to a dwelling unit?

Answer: _____

_____

_____

_____

Reference: _____

2. After you install a satellite dish at a local residence, the inspector informs you that you must bury the cable from the satellite dish to the house in accordance with Table 300.5 of the *NEC®*. Was that inspector correct?

Answer: _____

_____

_____

_____

Reference: _____

3. Does the *Code* require grounding the metal sheath of CATV cable entering a dwelling?

Answer: _____

_____

_____

_____

Reference: _____

4. According to the *NEC®*, open conductors for communication equipment on a building shall be separated at least _____ feet (_____ mm) from *lightning* conductors, where practicable.

Answer: _____

_____

_____

_____

Reference: _____

5. In a highly hazardous location where propane is used, the customer states that the wiring to the control circuit is of sufficient low voltage that a spark cannot be generated from this circuit. The devices and wiring have a control drawing number on them. What type wiring is most likely employed in this installation where there is not enough voltage to generate a spark sufficient to cause an explosion?

Answer: _____

_____

_____

_____

Reference: _____

6. You recently encountered a receptacle with an orange triangle on the face of it. You were told that this was an isolated ground receptacle. Where can this be found in the *NEC®*?

Answer: _____

_____

_____

_____

Reference: _____

7. In wiring a large church, an installation of a pipe organ is required. Are pipe organs covered in the *NEC®*?

Answer: _____

_____

_____

_____

Reference: _____

8. In making an installation, it was noted that the clearance between two large switchboards was rather tight, only about 3 feet (900 mm). With the doors open, there was not room to pass. Are there any requirements in the *NEC*® to cover the requirements for working clearances?

Answer: _____

_____

_____

_____

Reference: _____

9. After you completed wiring a new home, the inspector came and told you to cover the panelboard, that if the sheetrockers got their spackling material or paint on the panel, it was a code violation. Is this correct?

Answer: _____

_____

_____

_____

Reference: _____

10. You recently mounted a new service on a new mobile home on a privately owned lot. The inspector turned it down and said this was not permitted. Which section of the *NEC*® did you violate?

Answer: _____

_____

_____

_____

Reference: _____

11. What is an optional standby system?

Answer: _____

_____

_____

_____

Reference: _____

12. Is color-coding required in the *NEC*®? You know that the neutral conductor is required to be white and the grounding conductor is required to be green, but what about the ungrounded conductors?

Answer: _____

_____

_____

_____

Reference: _____

13. In making an installation to an outdoor ballfield, you proposed a common neutral to be run with multiple branch-circuit conductors. How many branch-circuit conductors can be supplied with a common neutral?

Answer: _____

_____

_____

_____

Reference: _____

14. You have been asked to wire a dairy barn. Are there any specific requirements in the *NEC*® for this type of installation?

Answer: _____

_____

_____

_____

Reference: _____

15. You recently observed a local electrician installing wiring under a carpet in a bank. Is this wiring method permitted? Which article covers wiring under carpets?

Answer: _____

_____

_____

_____

Reference: _____

16. More buildings are increasingly requiring lightning arrestors to be installed on the service. Which section of the *NEC*® covers this installation?

Answer: _____

_____

_____

_____

Reference: _____

17. Are temporary wiring methods covered in the *National Electrical Code*®? You regularly install a temporary meter loop to supply power to the construction crews until the permanent power system has been installed. Where is this covered in the *NEC*®?

Answer: _____

_____

_____

_____

Reference: _____

18. A 480-volt, 3-phase circuit with a rated load of 150 kW draws 210 amperes. The power factor is _____ percent.

    Answer: _____

    _____

    _____

    _____

    Reference: _____

19. A _____-horsepower, 208-volt, 3-phase motor operating at 91 percent efficiency has an input current of 141 amperes. (1 hp = 746 watts)

    Answer: _____

    _____

    _____

    _____

    Reference: _____

20. A feeder supplying a second building requires a maximum conductor resistance of 0.02 ohm. The minimum size conductor must have a resistance of no more than _____ ohm(s) per 1000 feet (305 m). The total length of the circuit conductor is 190 feet (58 m).

    Answer: _____

    _____

    _____

    _____

    Reference: _____

Chapter Fourteen

ELECTRICAL AND *NEC®* QUESTION REVIEW

Now you are about to begin the final chapter of this book. You should have a good understanding of the *Code* arrangement and its content. You should now understand that the chapters are arranged in the order in which you use them in most common installations, from Article 90—Introduction; Chapter 1—general requirements; Chapter 2—wiring and protection; Chapter 3—wiring methods and materials; Chapter 4 —general-use equipment; Chapters 5, 6, and 7—special occupancies, uses, and equipment; and Chapter 8 —communication systems. The final chapter in the *NEC®* is Chapter 9, which has the tables. Furthermore, you should now understand this arrangement and know the articles contained in each chapter. For instance, you should understand that the article covering rigid metal conduit (RMC) would automatically be found in Chapter 3, because rigid metal conduit is a wiring method. You should understand that the conditions for installation of a luminaire (light fixture) would be found in Chapter 4, because luminaires (light fixtures) are general-use equipment. Or you should understand that hospital installation requirements would be found in Chapter 5, because a hospital is a special occupancy and those are found in Chapter 5, such as mobile homes, hazardous materials and spray booths, commercial garages, and so on. You should automatically remember that a sign would be found in Chapter 6, special equipment, because although signs are common, they are not general equipment found in all types of installations. Signs are special equipment and are found in Chapter 6. These things must be understood before you take the examination. It is necessary so that you might use the *National Electrical Code®* efficiently as a reference document. Memorization is not recommended. However, with a good, clear understanding, one can refer to the Table of Contents in the front of the book and quickly find the article number if you have an understanding of where that article should be found in the *Code*. You should then be able to go to the Index and find the proper reference so that you might quickly look up the question and prove the answer. Once this has been done, you should have no problem passing any competency exam given throughout the country that is based on your knowledge of the *National Electrical Code®*.

With that competency achieved and your understanding of the *National Electrical Code®* arrangement ensured, the next important task you must improve is the understanding of what each question is asking. You must read the question through, then determine exactly what the question is asking. You will find key words, and with those key words you can simplify the question generally so that it can be quickly researched using the Table of Contents, the Index, and your knowledge of the *Code* arrangement.

Example: As an example, in the following question, select the key words and simplify the question:

"May Type NM cable be used for a 120/140-volt branch-circuit as temporary wiring in a building under construction where the cable is supported on insulators at intervals of not more than 10 feet (3 m)?"

Now, you may rephrase the question slightly different in your mind, or you may look for the key words. NM cable is a key word, temporary wiring is a key, and supported on insulators at intervals of not more than 10 feet (3 m) is a key. So,

"Is NM cable permitted as temporary wiring where supported on insulators?"

Now, with the three key words in mind, we must remember that they are not asking where this wiring method is acceptable, they are telling you that the installation would be temporary wiring. If we go to the Table of Contents in the front of the *NEC®*, we find that temporary wiring systems is found in Article 590. As we turn to Article 590 and we look in the scope under 590.1, we find out that, yes, lesser methods are acceptable. As we research that article quickly, we find in 590.4(C) that this open wiring cable is an acceptable method, provided that it is supported on insulators at intervals of not more than 10 feet (3 m), so the answer would be yes. References would be *NEC®* 590.1 and 590.4(C). As you see in this case when selecting the key words or phrases, using the Table of Contents was sufficient and it was not necessary to go to the Index.

Example: Let us look at another example. The question is:

"Is liquidtight flexible nonmetallic conduit (LFNC) permitted to be used in circuits in excess of 600 volts?"

As we look at that question, we see that liquidtight flexible nonmetallic conduit is the key phrase. Then we look at permitted use as a key phrase and 600 volts as the final key phrase.

Again, we can go to the Table of Contents and find that liquidtight flexible nonmetallic conduit is found in Article 356. As we turn to Article 356, we see uses permitted in 356.10, uses not permitted in 356.12. Immediately as we scan down the uses

permitted and the uses not permitted; we find in Section 356.12(4) that liquidtight flexible nonmetallic conduit is not permitted for circuits in excess of 600 volts. However, you will note that there is an exception for electrical signs, so the answer is: "Yes, for electrical signs by exception. No, generally."

If you will take each question and break it down so that you are looking at the key phrase or key word, you will find it much easier to reference the correct *Code* article and section, and you will not waste the valuable time needed to complete your examination. All questions will not be as easy as these two examples to identify the key phrases, but in all questions there are key phrases or key words, and with the practice you will receive during the following exercises in Chapter 14, you should greatly improve your competency and speed when preparing for an examination, which will allow you ample time to get through the examination and still have some time to do additional research on the tough questions. What you will find by following these examples is that there will be fewer tough questions in the exam. Good luck!

2005 Highlighted Code Changes

There were several thousand proposals submitted to change the 2002 *NEC®* and over 2,000 public comments. The *NEC®* Technical Correlating Committee continued with appointed task groups to make the *Code* more friendly and easier to interpret. Several new articles were added.

Articles New for the 2005 *NEC®*

Article 80 has been relocated as a new addendum G. There is a new **Article 353** for high density polyethylene (HDPE) conduit. **Article 366**—Auxiliary Gutters and **Article 368**—Busways have been rewritten and renumbered to conform to the numbering style in Chapter 3. A new **Article 409**—Industrial Control Panels has been added. A new **Article 506** is added to cover the Zone alternative system for hazardous dusts. Temporary Installations have been relocated in the 2005 *NEC®* and is renumbered as **Article 590**. A new **Article 682** has been added to cover bodies of water not covered by Article 680. This will include fish farms and similar bodies of water.

PRACTICE EXAMINATION

The following test was developed and administered by a nationally known seminar presenter and author at a meeting of *Code* authorities. Although this test is not representative of national electricians' tests, it does provide a good exercise in researching the *NEC*®. You will find the questions to be somewhat tricky in some cases. However, if you can pass this examination within one hour, you should be able to pass most national examinations. It will exercise your knowledge of the *NEC*®, your ability to use the document as a reference, and your ability to research variable difficult questions.

1. Equipment enclosed in a case or cabinet that is provided with a means of sealing or locking so that live parts cannot be made accessible without opening the enclosure is said to be
   A. guarded.
   B. protected.
   C. sealable equipment.
   D. lockable equipment.

   Answer: _____ Reference: _____

2. The letter(s) _____ indicate(s) two insulated conductors laid parallel within an outer nonmetallic covering.
   A. D
   B. M
   C. T
   D. II

   Answer: _____ Reference: _____

3. Class 1 circuit conductors shall be protected against overcurrent
   A. in accordance with the values specified in Table 310.16 through Table 310.31 for 14 AWG and larger.
   B. shall not exceed 7 amperes for 18 AWG.
   C. shall not exceed 10 amperes for 16 AWG.
   D. and derating factors do not apply.
   E. all of the above

   Answer: _____ Reference: _____

4. Enclosures for switches or circuit breakers shall not be used as
   A. junction boxes.
   B. auxiliary gutters.
   C. raceways.
   D. all of the above

   Answer: _____ Reference: _____

5. A plug fuse of the Edison base has a maximum rating of _____ amperes.
   A. 20
   B. 30
   C. 40
   D. 50
   E. 60

   Answer: _____ Reference: _____

6. Non–power-limited fire-protective signaling circuit conductors shall be
   A. solid copper.
   B. bunch-tinned stranded copper.
   C. bonded stranded copper.
   D. any of the above

   Answer: _____   Reference: _____

7. When conduit or tubing nipples having a maximum length not to exceed _____ inches (_____ mm) are installed between boxes and similar enclosures, the fill shall be permitted to 60 percent.
   A. 6 (150)
   B. 12 (300)
   C. 24 (600)
   D. 30 (750)

   Answer: _____   Reference: _____

8. Circuit breakers shall be
   A. capable of being opened by manual operation.
   B. capable of being closed by manual operation.
   C. trip free.
   D. all of the above
   E. A and B only

   Answer: _____   Reference: _____

9. For a circuit operating at less than 50 volts, standard lampholders having a rating not less than _____ watts shall be used.
   A. 300
   B. 550
   C. 660
   D. 770

   Answer: _____   Reference: _____

10. 18 AWG TFF wire is rated _____ amperes.
    A. 14
    B. 10
    C. 8
    D. 6

    Answer: _____   Reference: _____

11. A 10 AWG solid copper wire has a cross-section area of _____ square inches (_____ mm^2).
    A. .008 (5.26)
    B. 1.21 (3.984)
    C. .006 (4.25)
    D. .109 (70.41)

    Answer: _____   Reference: _____

12. Conductors of light or power can occupy the same enclosure or raceway with conductors of power-limited fire-protective signaling circuits.
    A. True
    B. False

    Answer: _____   Reference: _____

13. Explanatory material in the *National Electrical Code*® is in the form of
    A. footnotes.
    B. fine print notes.
    C. obelisks and asterisk.
    D. red print.

    Answer: _____   Reference: _____

14. Straight runs of trade size 1¼ (35) rigid metal conduit using threaded couplings can be secured at not more than _____ foot (_____ m) intervals.
    A. 5 (1.5)
    B. 10 (3)
    C. 12 (3.7)
    D. 14 (4.3)

    Answer: _____   Reference: _____

15. An isolating switch is one that
    A. is not readily accessible to persons unless special means for access are used.
    B. is capable for interrupting the maximum operating overload current of a motor.
    C. is intended for use in general distribution and branch-circuits.
    D. is intended for isolating an electrical circuit from the source of power.

    Answer: _____   Reference: _____

16. Where damage to remote-control circuits of safety-control equipment would produce a hazard, all conductors of this Class 1 circuit shall be installed in _____ or otherwise suitably protected from physical damage.
    I.   mineral-insulated or metal clad cable
    II.   rigid metallic conduit
    III. rigid nonmetallic conduit
    A. I only
    B. II only
    C. III only
    D. I, II, or III

    Answer: _____   Reference: _____

17. _____ on equipment to be grounded shall be removed from contact surfaces to assure good electrical continuity.
    A. Conductive coatings
    B. Nonconductive coatings
    C. Manufacturers instructions
    D. all of the above

    Answer: _____   Reference: _____

18. A _____ conductor is one having one or more layers of nonconducting materials that are not recognized by this *Code*.
    A. noninsulating
    B. bare
    C. covered
    D. none of these

    Answer: _____   Reference: _____

19. To guard live parts, _____ accessible to qualified persons only.
    A. isolate in a room
    B. locate on a balcony
    C. enclose in a cabinet
    D. any of these

    Answer: _____   Reference: _____

20. Conductors larger than 4/0 AWG are measured in
    A. inches.
    B. circular mils.
    C. square inches.
    D. none of the above

    Answer: _____   Reference: _____

21. Measurement of preference: The metric system used in the 2005 *NEC*® is known as the
    A. English metric system.
    B. International system of units.
    C. SOPT metric conversion.
    D. hard metric conversion.

    Answer: _____   Reference: _____

22. The voltage of a circuit is defined by the *Code* as the _____ root-mean-square (effective) difference of potential between any two conductors in the circuit.
    A. lowest
    B. greatest
    C. average
    D. nominal

    Answer: _____   Reference: _____

23. Doorbell wiring, rated as Class 2, in a residence _____ run in the same raceway with light and power conductors.
    A. is permitted with 600-volt insulation
    B. shall not be
    C. shall be
    D. is permitted if insulation is equal to highest installed

    Answer: _____   Reference: _____

24. Equipment or materials to which has been attached a symbol or other identifying mark acceptable to the authority having jurisdiction is known as
    A. listed.
    B. labeled.
    C. approved.
    C. rated.

    Answer: _____   Reference: _____

25. The grounded conductor of a branch-circuit shall be identified by a _____ color.
    I.   gray
    II. continuous green
    III. continuous white
    IV. green with yellow stripe
    A. I and III
    B. II and IV
    C. III only
    D. II only

    Answer: _____   Reference: _____

26. Circuits for lighting and power shall not be connected to any system containing
    A. hazardous material.
    B. trolley wires with ground returns.
    C. poor wiring methods.
    D. dangerous chemicals or gases.

    Answer: _____   Reference: _____

27. The minimum clearance between an electric space heating cable and an outlet box shall not be less than _____ inches (_____ mm).
    A. 8 (200)
    B. 12 (300)
    C. 18 (450)
    D. 6 (150)

    Answer: _____   Reference: _____

28. The ampacity requirements for a disconnecting means of X-ray equipment shall be based on _____ percent of the input required for the momentary rating of the equipment if greater than the long-time rating.
    A. 125
    B. 100
    C. 50
    D. none of the above

    Answer: _____   Reference: _____

29. Open conductors installed outside shall be separated from open conductors of other circuits by not less than _____ inches (_____ mm).
    A. 4 (100)
    B. 6 (150)
    C. 8 (200)
    D. 10 (250)
    E. 12 (300)

    Answer: _____   Reference: _____

30. The minimum headroom of working spaces about motor control centers shall be
    A. 3½ feet (1 m).
    B. 5 feet (1.5 m).
    C. 6 feet, 6 inches (2 m).
    D. 6 feet, 3 inches (1.8 m).

    Answer: _____   Reference: _____

31. Where fixed multioutlet assemblies are employed in locations where a number of appliances are likely to be used simultaneously, each foot or fraction thereof shall be considered as an outlet of not less than _____ VA.
    A. 180
    B. 200
    C. 60
    D. 35

    Answer: _____   Reference: _____

32. Soft-drawn or medium-drawn copper lead in conductors for television equipment antenna systems shall be permitted where the maximum span between points of support is less than _____ feet (_____ m).
    A. 35 (11)
    B. 30 (9.1)
    C. 20 (6)
    D. 10 (3)

    Answer: _____   Reference: _____

33. For a one-family dwelling, the service disconnect shall not be less than _____.
    A. 60
    B. 100
    C. 150
    D. 200

    Answer: _____   Reference: _____

34. Which of the following machines shall be provided with speed-limiting devices or other speed-limiting means?
    I. series motors
    II. induction
    III. self-excited DC motors
    A. I only
    B. II only
    C. III only
    D. I and III

    Answer: _____   Reference: _____

35. Where within _____ feet (_____ m) of any building or other structure, open wiring on insulators shall be insulated or covered.
    A. 4 (1.2)
    B. 3 (900 mm)
    C. 6 (1.8)
    D. 10 (3)

    Answer: _____   Reference: _____

36. High-voltage conductors in tunnels shall be installed in
    A. rigid metal conduit.
    B. MC cable.
    C. other metal raceways.
    D. any of the above

    Answer: _____   Reference: _____

37. For garages and outbuildings on residential property, a _____ suitable for use on branch-circuits shall be permitted as the disconnecting means.
    A. snap switch
    B. set of three-way or four-way snap switches
    C. set of two-way or four-way snap switches
    D. A or B
    E. none of the above

    Answer: _____   Reference: _____

38. Utilization equipment fastened in place connected to a branch-circuit with other loads shall not exceed _____ percent of the branch-circuit rating.
    A. 50
    B. 60
    C. 80
    D. 100

    Answer: _____   Reference: _____

39. The maximum amperage rating of a 4 inch × ½ inch busbar is _____ amperes.
    A. 500
    B. 1000
    C. 700
    D. 2000

    Answer: _____   Reference: _____

40. For temporary wiring over 600 volts, nominal, _____ shall be provided to prevent access of other than authorized and qualified personnel.
    I. fencing
    II. barriers
    III. signs
    A. I only
    B. II only
    C. III only
    D. I or II

    Answer: _____   Reference: _____

41. Each continuous-duty motor _____ horsepower or less, not permanently installed, is nonautomatically started, and is within sight of the controller shall be permitted to be protected against overload by the branch-circuit protective device.
    A. ⅛
    B. ½
    C. ¾
    D. 1

    Answer: _____   Reference: _____

42. The neutral (grounded) conductor in a mobile home shall be insulated from the equipment grounding system to which of the following location(s)?
    A. range and oven
    B. distribution panel
    C. clothes dryer
    D. all of the above

    Answer: _____   Reference: _____

43. The bottom of sign and outline lighting enclosures shall not be less than _____ feet (_____ m) above areas accessible to vehicles.
    A. 12 (3.7)
    B. 14 (4.25)
    C. 16 (4.9)
    D. 18 (5.5)

    Answer: _____   Reference: _____

44. The conductor used to ground the outer cover of a coaxial cable shall be
    I.   insulated.
    II. 14 AWG minimum.
    III. guarded from physical damage when necessary.
    A. I only
    B. II only
    C. III only
    D. I, II, III

    Answer: _____   Reference: _____

45. Where energized live parts are exposed, the minimum clear workspace shall not be less than _____ feet (_____ m) high for over 600 volts.
    A. 3 (900 mm)
    B. 5 (1.8)
    C. 3½ (1)
    D. 6¼ (1.9)
    E. 6½ (2)

    Answer: _____   Reference: _____

46. Conductors _____ AWG or larger supported on solid knobs shall be securely tied thereto by tie wires having an insulation equivalent to that of the open wire.
    A. 14
    B. 12
    C. 10
    D. 8
    E. 6

    Answer: _____   Reference: _____

47. A branch-circuit that supplies a number of outlets for lighting and appliances is known as a
    A. general-purpose branch-circuit.
    B. a multipurpose branch-circuit.
    C. a utility branch-circuit.
    D. none of the above

    Answer: _____   Reference: _____

48. The minimum size conductor that can be used for an overhead feeder from a residence to a remote garage is
    A. 10 AWG cu.
    B. 12 AWG cu.
    C. 6 AWG al.
    D. 10 AWG al.

    Answer: _____   Reference: _____

49. Two or more _____, 120-volt small-appliance branch-circuits 15- or 20-ampere receptacle outlets in dwelling unit, kitchen, dining room, breakfast room, pantry, or similar dining areas are required in all dwelling units.
    A. 15-ampere
    B. 20-ampere
    C. 30-ampere
    D. 15- or 20-ampere

    Answer: _____ Reference: _____

50. Each kitchen countertop surface in a dwelling unit shall be supplied by not fewer than _____ small-appliance branch-circuit(s).
    A. one
    B. two
    C. three
    D. no minimum

    Answer: _____ Reference: _____

2005 *CODE* CHANGES QUIZ 1

> Each lesson is designed purposely to require the student to apply the entire *NEC®* text, and not specific chapters or articles. It has been found that when studying for a timed, open-book examination, the student must gain proficiency in the Table of Contents, the Index, and the ability to move quickly from cover to cover to find the correct answer to each question in a timely fashion.

1. Which article in the *NEC®* is applicable when correctly installing the electrical systems for commercial fish ponds and sewer ponds?

   Answer: _____ Reference: _____

2. High-density polyethylene conduit as an assembly with conductors enclosed. Where is high-density polyethylene conduit covered in the *Code*?

   Answer: _____ Reference: _____

3. Where are the requirements for installing temporary wiring on a construction job for the workers only during construction?

   Answer: _____ Reference: _____

4. An area including a basin with one or more of the following—a toilet, a tub, or a shower—is defined in the *NEC®* as a _____.

   Answer: _____ Reference: _____

5. Hazardous locations determined to be neither Class I, Division 1; Class I, Division 2; Class I, Zone 0; Class I, Zone 1; Class I, Zone 2; Class II, Division 1; Class II, Division 2; Class III, Division 1; Class III, Division 2; nor any combination thereof are _____.

   Answer: _____ Reference: _____

6. The 2002 *NEC®* Article 80 contained administration and enforcement rules. Was this material deleted from the 2005 *NEC®*?

   Answer: _____ Reference: _____

7. The main bonding jumper is used to bond the grounded conductor to the grounding electrode and equipment grounding. What is the system bonding jumper used for?

   Answer: _____ Reference: _____

8. The conductor used to connect the grounding electrode(s) to the equipment grounding conductor, to the grounded conductor, or to both, at the service, at each building or structure where supplied by a feeder(s) or branch-circuit(s) from a common service, or at the source of a separately derived system is called the _____.

   Answer: _____ Reference: _____

9. An enclosure identified for use in underground systems, provided with an open or closed bottom, and sized to allow personnel to reach into, but not enter, for the purpose of installing, operating, or maintaining equipment or wiring or both is defined as a _____.

   Answer: _____   Reference: _____

10. A kitchen not located within a dwelling is defined as an area with a sink and permanent facilities for food preparation, and cooking. What section of the *Code* places special requirements for these areas?

    Answer: _____   Reference: _____

11. All 125-volt, 15- or 20-ampere receptacles located in other than dwelling units outdoors in public places must have _____ for personnel.

    Answer: _____   Reference: _____

12. Where is the allowable minimum bending radius for all conduit and tubing found?

    Answer: _____   Reference: _____

13. Article 830 covers network-powered broadband communications systems that are covered by the *NEC*®. Those not covered are found in Section _____ .

    Answer: _____   Reference: _____

14. Article 820 covers CATV systems and radio distribution systems. How is the maximum number of coaxial cables permitted in conduit calculated?

    Answer: _____   Reference: _____

15. GFCI protection is required for 15A and 20A, 125V, receptacles in aircraft hangers. True or False?

    Answer: _____   Reference: _____

2005 *CODE* CHANGES QUIZ 2

1. Abandoned power cables, communication cables, and connecting and interconnection cables are required to be removed under raised floors in information technology (IT) rooms unless contained in _____.

   Answer: _____   Reference: _____

2. Contact conductors and exposed short lengths of conductors at resistors, collectors, and other equipment for cranes or hoists are not required to be _____.

   Answer: _____   Reference: _____

3. Outdoor or under-chassis line-voltage (120 volts, nominal, or higher) wiring exposed to moisture or physical damage shall be protected by _____ or _____ unless they are routed closely against the frame and equipment enclosure of mobile or manufactured homes.

   Answer: _____   Reference: _____

4. A service for a carnival would have to comply with all the requirements of Article _____ and Sections _____.

Answer: _____   Reference: _____

5. The wiring methods for Zone 20 are found in Section _____.

Answer: _____   Reference: _____

6. Resistors commonly found in electrical control cabinets and other electrical apparatus are covered in _____.

Answer: _____   Reference: _____

7. The disconnecting means for motor circuits rated 600 volts, nominal, or less shall have an ampere rating not less than _____ percent of the full-load current rating of the motor, except for specific listed nonfused motor-circuit switches.

Answer: _____   Reference: _____

8. Receptacles shall not be installed within or directly over _____ or _____.

Answer: _____   Reference: _____

9. Conductors in vertical poles used as raceways shall be supported per Section _____.

Answer: _____   Reference: _____

10. Flexible cords and cables must conform to the description provided in _____.

Answer: _____   Reference: _____

11. Open wiring on insulators shall not be permitted in other than industrial or agricultural installations and where concealed by the building structure or where the voltage exceeds _____.

Answer: _____   Reference: _____

12. Do the ampacity adjustment factors apply to conductors in cellular metal raceways?

Answer: _____   Reference: _____

13. Nonmetallic cable tray shall be permitted only in corrosive areas and in areas requiring voltage isolation. True or False?

Answer: _____   Reference: _____

14. Busways may not be installed where subject to _____ or severe _____.

Answer: _____   Reference: _____

15. If I ask for 10 lengths of 27 IMC, what should I receive from the supplier?

Answer: _____   Reference: _____

2005 *CODE* CHANGES QUIZ 3

1. Integrated gas spacer cable is not permitted to be used as interior wiring, exposed or in contact with buildings or _____.

    Answer: _____   Reference: _____

2. _____ receptacles or covers are required in psychiatric and pediatric general-care areas where accessible to patients.

    Answer: _____   Reference: _____

3. The _____ cannot be used as the sole equipment grounding conductor or as an effective ground-fault current path.

    Answer: _____   Reference: _____

4. The processes of purging or pressurizing supplying enclosures in Class 1 hazardous locations to an acceptable level are acceptable protection techniques. What NFPA standard covers these two protection techniques?

    Answer: _____   Reference: _____

5. Single conductors specified in Table 310.13 must only be installed in a recognized wiring method of Chapter 3 or be part of a Chapter 3 wiring method except in accordance with _____.

    Answer: _____   Reference: _____

6. Where grounded conductors of different systems are installed in the same raceway, cable, box, auxiliary gutter, or other type of enclosure, each grounded conductor shall be identified by system identification that distinguishes each system. This means of identification shall be permanently posted at each _____.

    Answer: _____   Reference: _____

7. An attachment plug and multioutlet receptacle assembly where used as permitted assembly shall consist of an attachment plug, a flexible cord, and a multioutlet receptacle strip. Each assembly must be energized from a receptacle outlet. The assembly must be listed, and the total length of the assembly shall not exceed _____ feet?

    Answer: _____   Reference: _____

8. In mobile or manufactured homes, receptacle outlets shall not be permitted to be installed within a _____ or _____ or _____ position in any countertop or _____ electric baseboard heaters, unless provided for in the listing or manufacturer's instructions.

    Answer: _____   Reference: _____

9. The conduit or tubing not smaller than _____ trade size for neon secondary circuit conductors, over 1000 volts, nominal, shall contain no more than _____ conductor.

Answer: _____   Reference: _____

10. Which article covers the requirements for electric cutting process equipment?

Answer: _____   Reference: _____

11. Is a disconnect switch required to be inside of the motor? Or can the disconnect be located elsewhere if it can be locked in the off position?

Answer: _____   Reference: _____

12. How is the size of an equipment bonding jumper determined?

Answer: _____   Reference: _____

13. Do raceways to be installed by directional boring equipment have to be listed for the purpose?

Answer: _____   Reference: _____

14. What cables do not have to be marked as "sunlight resistant" when they are used in installations exposed to direct sunlight?

Answer: _____   Reference: _____

15. Is it permissible to have 220-volt lighting circuits in a dwelling?

Answer: _____   Reference: _____

16. Would a 120-volt receptacle installed outdoors at a dwelling for a sewer lift pump require GFCI protection?

Answer: _____   Reference: _____

17. What is the trade size of a 103 intermediate conduit?

Answer: _____   Reference: _____

18. Is a green or bare equipment grounding conductor required to be installed with the circuit conductors installed in electrical metallic tubing?

Answer: _____   Reference: _____

19. Where are the installation requirements for industrial control cabinets found?

Answer: _____   Reference: _____

20. The TVSS shall be connected to each _____ conductor.

Answer: _____   Reference: _____

2005 *CODE* CHANGES QUIZ 4

1. Is it permissible for a switch in a patient care area, located outside the patient vicinity, to be supplied by a switch loop consisting of Type MC cable?

   Answer: _____

   _____

   _____

   _____

   Reference: _____

2. What is the difference between a lighting fixture and a luminaire?

   Answer: _____

   _____

   _____

   _____

   Reference: _____

3. Does the 2005 *NEC*® contain adoptive ordinance language similar to other model codes?

   Answer: _____

   _____

   _____

   _____

   Reference: _____

4. Has the 2005 *Code* been changed or modified in any way to make it a more suitable document for international use?

   Answer: _____

   _____

   _____

   _____

   Reference: _____

5. Does 90.4 address signaling and communications systems as part of the term "electrical installations"?

   Answer: _____

   _____

   _____

   _____

   Reference: _____

6. Are informative product standards referenced in the *NEC*®? If so, where?

   Answer: _____

   _____

   _____

   _____

   Reference: _____

7. Are metric measurements the measurement system of preference in the 2005 *Code*?

   Answer: _____

   _____

   _____

   _____

   Reference: _____

8. Where are the requirements for receptacles, cord connectors, and attachment plugs found in the 2005 *NEC*®?

   Answer: _____

   _____

   _____

   _____

   Reference: _____

9. Would the definition of a garage in Article 100 include a building where electric vehicles are stored?

   Answer: _____

   _____

   _____

   _____

   Reference: _____

10. Would 110.22 literally not require identification of a disconnecting means that was not required by the *Code*?

Answer: _____

_____

_____

_____

Reference: _____

11. Is panic hardware on electrical room doors required?

Answer: _____

_____

_____

_____

Reference: _____

12. Where can the requirements for the specific wiring methods for electric discharge luminaires (lighting fixtures) supported independent of the outlet box be found?

Answer: _____

_____

_____

_____

Reference: _____

13. What does the new term "luminaire" used throughout the *Code* refer to?

Answer: _____

_____

_____

_____

Reference: _____

14. Does the *Code* prohibit receptacles from being installed in a face-up position in a dwelling-unit laundry room countertop?

Answer: _____

_____

_____

_____

Reference: _____

15. Are dwelling unit bedroom lighting circuits required to be AFCI protected?

Answer: _____

_____

_____

_____

Reference: _____

16. Receptacles are usually installed in appliance garages on dwelling-unit kitchen countertops. Can this receptacle be considered as one of the required receptacles for the countertop?

Answer: _____

_____

_____

_____

Reference: _____

17. Is GFCI protection required for electrically heated floors in bathrooms, hydromassage bathtub, spa, and hot tub locations?

Answer: _____

_____

_____

_____

Reference: _____

18. Is it permissible to use a gray-colored conductor as a grounded (neutral) conductor?

Answer: _____

_____

_____

_____

Reference: _____

19. Are branch-circuits over 600 volts permitted in residential and commercial occupancies where there are no maintenance or supervision provisions?

Answer: _____

_____

_____

_____

Reference: _____

20. Is a service conductor for an overhead service drop permitted to be bare?

Answer: _____

_____

_____

_____

Reference: _____

2005 *CODE* CHANGES QUIZ 5

1. *NEC*® 230.46 allows service-entrance conductors, including directly buried conductors, to be spliced. Are service-lateral conductors permitted to be to be spliced?

Answer: _____

_____

_____

_____

Reference: _____

2. Is cable tray a wiring method for all service conductors?

Answer: _____

_____

_____

_____

Reference: _____

3. *NEC*® 230.46 permits service-entrance conductors to be spliced using clamped or bolted connections. Could a crimp connection or exothermic welding be used for this purpose?

Answer: _____

_____

_____

_____

Reference: _____

4. Would 230.70 allow the use of a shunt trip switch at the location specified in 230.70(A)(1) as the service disconnecting means if the circuit breaker were located inside the building?

Answer: _____

_____

_____

_____

Reference: _____

5. A mobile home is equipped with a chassis bonding connection as required in 550.16(C)(1). The metal gas piping is bonded with a factory-installed bonding jumper connected to the mobile home frame at a point remote from the chassis bonding connection. Does this installation meet all the requirements?

Answer: _____

_____

_____

_____

Reference: _____

6. Does the *Code* specifically allow a molded case switch to be used as a motor controller?

Answer: _____

_____

_____

_____

Reference: _____

7. The service disconnecting means is located outside a building. Can it be located 50 feet (15 m) away or within sight of the building?

Answer: _____

_____

_____

_____

Reference: _____

8. Where can requirements for separately derived systems that operate at 120 volts line-to-line and 60 volts line-to-ground be found?

Answer: _____

_____

_____

_____

Reference: _____

9. What section of the *Code* permits solar photovoltaic systems, fuel cell systems, or interconnected electric power production sources to be connected to the supply-side of the service disconnect?

Answer: _____

_____

_____

_____

Reference: _____

10. There is presently no maximum rating or setting for overcurrent devices protecting fire pumps and related equipment. How would overcurrent devices for fire pumps be sized?

    Answer: _____

    _____

    _____

    _____

    Reference: _____

11. Is ground-fault circuit-interrupter protection required for all single-phase, 15- and 20-ampere receptacles in commercial kitchens?

    Answer: _____

    _____

    _____

    _____

    Reference: _____

12. Where in the *Code* are the branch-circuit overcurrent devices supplying a motel room required to be readily accessible to the motel room occupant?

    Answer: _____

    _____

    _____

    _____

    Reference: _____

13. Are circuit breakers permitted to be used as a switch? Are they permitted for switching HID luminaires (lighting fixtures) if they are only marked SWD?

    Answer: _____

    _____

    _____

    _____

    Reference: _____

14. Are there presently any *Code* requirements for the removal or termination of abandoned low-voltage cables in buildings or structures?

    Answer: _____

    _____

    _____

    _____

    Reference: _____

15. Where in the *Code* does it require that temporary electrical power and lighting installations for holiday decorative lighting be limited to 90 days?

Answer: _____

_____

_____

_____

Reference: _____

16. Are underground water pipe supplementary grounding electrodes required to meet the requirements of 250.56 if they are rod pipe or plate types of electrodes?

Answer: _____

_____

_____

_____

Reference: _____

17. Is it permissible to connect a grounding electrode to equipment such as parking light poles?

Answer: _____

_____

_____

_____

Reference: _____

18. What is the difference in performance between a surge arrester and a surge suppressor?

Answer: _____

_____

_____

_____

Reference: _____

19. In walls and ceilings constructed of wood or other combustible surface material, boxes shall be flush with the finished surface or project therefrom. Is gypsum board (sheetrock) considered a combustible surface material?

Answer: _____

_____

_____

_____

Reference: _____

20. A luminaire (lighting fixture) that weighs more than _____ shall be supported independently of the outlet box unless the outlet box is listed for the weight to be supported.

Answer: _____

_____

_____

_____

Reference: _____

2005 *CODE* CHANGES QUIZ 6

1. Type AC cable shall be permitted to be unsupported where the cable is not more than _____ from the last point of support for connections within an accessible ceiling to luminaire(s) (lighting fixture[s]) or equipment.

Answer: _____

_____

_____

_____

Reference: _____

2. Does the *Code* require a raceway to be sealed where it passes from the interior to the exterior of a building where condensation is known to be a problem?

Answer: _____

_____

_____

_____

Reference: _____

3. Does the *Code* allow cables to be used as a means of support for other cables?

Answer: _____

_____

_____

_____

Reference: _____

4. Is liquidtight flexible metal conduit permitted to be used in ducts, plenums, and other air-handling spaces as long as no one run exceeds 6 feet (1.8 m) in length?

Answer: _____

_____

_____

_____

Reference: _____

5. Which article of the *Code* covers temporary installations?

Answer: _____

_____

_____

_____

Reference: _____

6. What is an arc-fault circuit interrupter as per *Code*?

Answer: _____

_____

_____

_____

Reference: _____

7. Do the ampacity adjustment factors for cables apply to all Type MC or AC cable?

Answer: _____

_____

_____

_____

Reference: _____

8. Is the messenger-supported wiring method permitted where subject to physical damage?

Answer: _____

_____

_____

_____

Reference: _____

9. Does the *Code* allow Type AC cable to be run unsupported in lengths not exceeding 6 feet (1.8 m) between luminaires (light fixtures) in accessible ceilings?

Answer: _____

_____

_____

_____

Reference: _____

10. New hybrid-type vehicles incorporate gasoline engines and electric motors. Would Article 625 or Article 511 govern repair facilities for these vehicles?

    Answer: _____

    _____

    _____

    _____

    Reference: _____

11. Could intermediate metal conduit (IMC) be used as an exposed vertical riser to support a lighted display case in a department store?

    Answer: _____

    _____

    _____

    _____

    Reference: _____

12. Does the *Code* permit threadless connectors to be used on the threaded end of rigid metal conduit (RMC)?

    Answer: _____

    _____

    _____

    _____

    Reference: _____

13. Are flat cable assemblies Type FC cable permitted to be installed where subject to physical damage?

    Answer: _____

    _____

    _____

    _____

    Reference: _____

14. Does the minimum wire bending space at terminals in Table 312.6(B) apply to aluminum conductors?

    Answer: _____

    _____

    _____

    _____

    Reference: _____

15. Would the metal faceplate (cover) grounding requirements apply to fan speed controls or spring-wound timers?

    Answer: _____

    _____

    _____

    _____

    Reference: _____

16. Are flash protection warning markings required for switchboards, panelboards, and motor control centers that are likely to require servicing while energized?

    Answer: _____

    _____

    _____

    _____

    Reference: _____

17. Is nonmetallic-sheathed cable Type NM permitted as open runs in dropped or suspended ceilings in other than one- and two-family and multifamily dwellings?

    Answer: _____

    _____

    _____

    _____

    Reference: _____

18. A service-drop point of attachment is anchored to a trade size 2 (53) rigid metal conduit riser 3 feet (900 mm) above the roofline. There is a threaded coupling on the riser between the highest strap and the weatherhead. Is this a *Code* violation?

    Answer: _____

    _____

    _____

    _____

    Reference: _____

19. Could the required emergency disconnect for gasoline dispensers serve as the maintenance disconnect required in _____?

    Answer: _____

    _____

    _____

    _____

    Reference: _____

20. Do the dedicated space requirements for switchboards and panelboards now apply to a fusible switch as well?

Answer: _____

_____

_____

_____

Reference: _____

2005 *CODE* CHANGE QUIZ 7

1. Does the AFCI branch-circuit protection for bedroom circuits require the use of AFCI circuit breakers?

Answer: _____

_____

_____

_____

Reference: _____

2. Could an individual refrigerator branch-circuit be considered in the small appliance branch-circuit calculation? Or would this individual circuit be included as part of the general lighting load?

Answer: _____

_____

_____

_____

Reference: _____

3. Why is Type AC cable not permitted to be used as a service-entrance wiring method?

Answer: _____

_____

_____

_____

Reference: _____

4. Would the bonding jumper connection to metal gas piping be required to be accessible after installation?

Answer: _____

_____

_____

_____

Reference: _____

5. Could Type NM cable be run unsupported in lengths not exceeding 4½ feet (1.4 m) between luminaires (lighting fixtures) in an accessible ceiling?

Answer: _____

_____

_____

_____

Reference: _____

6. Is nonferrous intermediate metal conduit (IMC) manufactured?

Answer: _____

_____

_____

_____

Reference: _____

7. Is electrical nonmetallic tubing (ENT) permitted outside exposed to the direct rays of the sun?

Answer: _____

_____

_____

_____

Reference: _____

8. Does the *Code* allow nonmetallic-sheathed cable (NM) to be installed in a complete raceway system?

Answer: _____

_____

_____

_____

Reference: _____

9. Is there a maximum number of bends allowed between pull points—for example, conduit bodies and boxes for liquidtight flexible metal conduit?

Answer: _____

_____

_____

_____

Reference: _____

10. Would a raceway, used as physical protection for direct burial conductors of over 600 volts emerging from the ground, be required to be listed for the purpose?

    Answer: _____

    _____

    _____

    _____

    Reference: _____

11. Luminaires (lighting fixtures) shall not be used as a raceway for circuit conductors unless _____ for use as a raceway.

    Answer: _____

    _____

    _____

    _____

    Reference: _____

12. What is Type NUCC? Is it required to be listed? Are the conductors inside NUCC required to be listed? Is the raceway required to be listed?

    Answer: _____

    _____

    _____

    _____

    Reference: _____

13. Is Type ENT permitted where subject to physical damage?

    Answer: _____

    _____

    _____

    _____

    Reference: _____

14. Cable trays shall be supported at intervals _____.

    Answer: _____

    _____

    _____

    _____

    Reference: _____

15. What is the maximum water level for a swimming pool?

Answer: _____

_____

_____

_____

Reference: _____

16. Is electrical metallic tubing (EMT) an equipment grounding conductor in accordance with the *Code* or is a separate equipment grounding conductor required?

Answer: _____

_____

_____

_____

Reference: _____

17. Do Tables 352.44(A) and (B) take into account heating of PVC conduit by direct sunlight, or are the tables only based on ambient temperature?

Answer: _____

_____

_____

_____

Reference: _____

18. Are nonmetallic wireways required to be listed?

Answer: _____

_____

_____

_____

Reference: _____

19. A service riser contains a 6 AWG bare neutral conductor. Is this permitted by the *Code*?

Answer: _____

_____

_____

_____

Reference: _____

20. Where in the *Code* does it require that voltage drop for branch-circuits not exceed 1.5 percent?

Answer: _____

_____

_____

_____

Reference: _____

2005 *CODE* CHANGE QUIZ 8

1. Is liquidtight flexible nonmetallic conduit permitted to enclose conductors above 600 volts?

Answer: _____

_____

_____

_____

Reference: _____

2. The equipment bonding jumper for liquidtight flexible nonmetallic conduit must be installed in accordance with _____.

Answer: _____

_____

_____

_____

Reference: _____

3. Are metal wireways required to be listed?

Answer: _____

_____

_____

_____

Reference: _____

4. The ampacity of the conductors from the generator terminals to the first distribution device(s) containing overcurrent protection shall not be less than _____ of the nameplate current rating of the generator.

Answer: _____

_____

_____

_____

Reference: _____

5. Single-phase cord-and-plug connected room air conditioners shall be provided with factory-installed _____ protection. The protection shall be an integral part of the attachment plug or be located in the power supply cord within 12 inches (300 mm) of the attachment plug.

   Answer: _____

   _____

   _____

   _____

   Reference: _____

6. Is each motor circuit required to include coordinated protection to automatically interrupt overload and fault currents in the motor, the motor-circuit conductors, and the motor control apparatus?

   Answer: _____

   _____

   _____

   _____

   Reference: _____

7. Unless protected by ground-fault circuit-interrupter protection for personnel, the secondary winding of the isolation transformer connected to the impedance heating elements shall not have an output voltage greater than _____ volts ac.

   Answer: _____

   _____

   _____

   _____

   Reference: _____

8. Are snap switches and control switches required to be grounded?

   Answer: _____

   _____

   _____

   _____

   Reference: _____

9. If a motor were accessible from a raised platform, would the *Code* allow the disconnect to be mounted higher than 6 feet, 7 inches (2 m) above the platform?

   Answer: _____

   _____

   _____

   _____

   Reference: _____

10. Is it acceptable to install more than one neutral under a terminal in a panelboard?

    Answer: _____

    _____

    _____

    _____

    Reference: _____

11. Exposed vertical risers from industrial machinery or fixed equipment shall be permitted to be supported at intervals not exceeding 20 feet (6 m). If the conduit is made up with threaded couplings, the conduit is firmly supported at the top and bottom of the riser, and no other means of intermediate support is readily available. What types of conduit are permitted to be used?

    Answer: _____

    _____

    _____

    _____

    Reference: _____

12. May a 25-ampere branch-circuit supply two or more outlets for fixed electric space-heating equipment?

    Answer: _____

    _____

    _____

    _____

    Reference: _____

13. The essential electrical distribution system for other than health care facilities shall be a _____ system.

    Answer: _____

    _____

    _____

    _____

    Reference: _____

14. Where Class II, Group E dusts are present in hazardous quantities, there are only _____ locations.

    Answer: _____

    _____

    _____

    _____

    Reference: _____

15. Overcurrent protection shall not be required for conductors from a battery rated less than _____ if the battery provides power for starting, ignition, or control of prime movers.

Answer: _____

_____

_____

_____

Reference: _____

16. All circuits and circuit modifications shall be legibly identified as to _____ or _____ on a circuit directory located on the face or inside of the panel door in the case of a panelboard, and at each switch on a switchboard.

Answer: _____

_____

_____

_____

Reference: _____

17. As presently worded, does 430.245(B) require a motor lead to be a solid conductor if larger than a 10 AWG ?

Answer: _____

_____

_____

_____

Reference: _____

18. A ballast in a fluorescent luminaire (lighting fixture) that is used for egress lighting and is energized only during an emergency shall not have _____.

Answer: _____

_____

_____

_____

Reference: _____

19. Would the *Code* allow the specified disconnecting means to be mounted on a removable cover of air-conditioning or refrigeration equipment if the disconnect were supplied from a flexible wiring method?

Answer: _____

_____

_____

_____

Reference: _____

20. A separable connector or a plug-and-receptacle combination in the supply line to an oven or cooking unit shall be approved for the _____ of the space in which it is located.

Answer: _____

_____

_____

_____

Reference: _____

2005 *CODE* CHANGES QUIZ 9

1. Can the electric "service" supplying a building or structure be supplied from a private generator?

Answer: _____

_____

_____

_____

Reference: _____

2. Where are the requirements for tunnel installations over 600 volts covered?

Answer: _____

_____

_____

_____

Reference: _____

3. A dedicated space equal to the depth and width of the equipment from the floor to a height of _____ or to the structural ceiling whichever is _____.

Answer: _____

_____

_____

_____

Reference: _____

4. Grounding and bonding requirements for tunnel installations over 600 volts are covered in *NEC®* _____.

Answer: _____

_____

_____

_____

Reference: _____

5. Generally, grounded conductors 6 AWG or smaller shall be identified by a continuous white or gray finish or by _____ on other than green insulation along its entire length.

Answer: _____

_____

_____

_____

Reference: _____

6. Is a receptacle installed in an accessory building such as a storage building at grade level associated with a dwelling required to have a ground-fault circuit-interrupter protection for personnel?

Answer: _____

_____

_____

_____

Reference: _____

7. What is an arc-fault circuit interrupter?

Answer: _____

_____

_____

_____

Reference: _____

8. May the service conductors originate at a generator where there is no utility serving the premises?

Answer: _____

_____

_____

_____

Reference: _____

9. Other electrical equipment associated with the electrical installation located above or below shall not be allowed to extend beyond the front of electrical equipment. True or False?

Answer: _____

_____

_____

_____

Reference: _____

10. The requirements for dedicated electrical space are located in Article _____.

    Answer: _____

    _____

    _____

    _____

    Reference: _____

11. Equipment requirements for over 600 volts nominal are located in Article _____.

    Answer: _____

    _____

    _____

    _____

    Reference: _____

12. An insulated grounded conductor of 6 AWG or smaller shall be identified by a continuous white or gray or _____.

    Answer: _____

    _____

    _____

    _____

    Reference: _____

13. Receptacles located in garages or accessory buildings having a floor located at or below grade and are limited to storage areas, work areas, or areas of similar use are required to be of the _____ type.

    Answer: _____

    _____

    _____

    _____

    Reference: _____

14. All branch-circuits supplying 125-volt, single-phase, 15- and 20-ampere branch-circuits installed in bedrooms shall be protected by _____.

    Answer: _____

    _____

    _____

    _____

    Reference: _____

15. Are the required two small-appliance circuits permitted to serve more than one kitchen?

Answer: _____

_____

_____

_____

Reference: _____

16. In dwelling units, at least one receptacle shall be located not more than _____ feet (_____ m) of the outside edge of each basin in bathrooms.

Answer: _____

_____

_____

_____

Reference: _____

17. In guest rooms of hotels, motels, and similar occupancies, at least _____ of the receptacles required by 210.52(A) shall be _____.

Answer: _____

_____

_____

_____

Reference: _____

18. An additional load of _____ shall be included for each 2 feet (600 mm) of track lighting or fraction thereof in all occupancies except dwelling units or guest rooms of hotels and motels.

Answer: _____

_____

_____

_____

Reference: _____

19. Outside feeders and branch-circuits not exceeding 300 volts-to-ground over residential property and driveways, and commercial areas not subject to truck traffic, shall have a minimum clearance from ground of _____ feet (_____ m).

Answer: _____

_____

_____

_____

Reference: _____

20. For a one-family dwelling, the feeder disconnecting means shall have a rating of not less than _____, 3-wire.

Answer: _____

_____

_____

_____

Reference: _____

2005 *CODE* CHANGES QUIZ 10

1. Which section of the *Code* permits one additional set of service-entrance conductors to supply a separate meter for public or common area loads not permitted to be supplied from an individual dwelling unit?

Answer: _____

_____

_____

_____

Reference: _____

2. For a one-family dwelling, the service disconnecting means shall have a rating of not less than _____, 3-wire.

Answer: _____

_____

_____

_____

Reference: _____

3. Do the overcurrent protection requirements in 240.4(D) for small conductors apply to motor or motor-operated appliances?

Answer: _____

_____

_____

_____

Reference: _____

4. Tap conductor is defined in *NEC*® _____.

Answer: _____

_____

_____

_____

Reference: _____

5. Where a tap is located outdoors except for the point of termination, the conductors are protected from physical damage and terminate in a single overcurrent device, the disconnecting means is readily accessible either outside or inside the building nearest the point of entrance, and the overcurrent device is integral or immediately adjacent to the disconnect, the length shall not exceed _____ feet (_____ m).

Answer: _____

_____

_____

_____

Reference: _____

6. Where in the *Code* is a "Supervised Industrial Installation" defined?

Answer: _____

_____

_____

_____

Reference: _____

7. Why is GFCI protection for personnel using 120-volt, 15- and 20-ampere rated receptacles, especially outdoors, not required for carnival and exhibition shows?

Answer: _____

_____

_____

_____

Reference: _____

8. I have been told that a tap conductor could not be tapped from a conductor that had been tapped from another conductor. Where is this stated in the *NEC®*?

Answer: _____

_____

_____

_____

Reference: _____

9. An individual 15-ampere branch-circuit is installed for refrigeration equipment in the kitchen of a dwelling unit per 210.52(B)(1), Exception No. 2. Does *NEC®* 220.52(A) require additional 1500 volt-amperes for this 15-ampere circuit?

Answer: _____

_____

_____

_____

Reference: _____

10. Do all the receptacle outlets in a hotel or motel room located conveniently for permanent furniture layout have to be accessible?

Answer: _____

_____

_____

_____

Reference: _____

11. If outlets end up behind nonstationary furniture such as a recliner chair, bed table, or desk such that they would have to be moved in order to plug into a receptacle outlet, does this mean they are not considered accessible?

Answer: _____

_____

_____

_____

Reference: _____

12. When should a transfer-switch switch the grounded conductor as well as the ungrounded conductors between a utility service and a generator service?

Answer: _____

_____

_____

_____

Reference: _____

13. Can a receptacle be located below the rim of a pedestal sink in a bathroom, and if so, how far below before it would not be considered as the required receptacle outlet for the bathroom basin?

Answer: _____

_____

_____

_____

Reference: _____

14. The requirement for a single 20-ampere circuit to supply the required receptacles in dwelling bathrooms does not permit any other loads to be connected to this circuit.

Answer: _____

_____

_____

_____

Reference: _____

15. Many large assembly and exhibition halls utilize cable trays for distribution of temporary wiring for an event. Is it permissible to use SO or SJT flexible cords in these cable trays for such temporary installations?

Answer: _____

_____

_____

_____

Reference: _____

16. The *Code* requires that electric motors used on fire pumps be _____ for fire pump service.

Answer: _____

_____

_____

_____

Reference: _____

17. Do the equipotential plane requirements of agriculture buildings apply to chicken houses?

Answer: _____

_____

_____

_____

Reference: _____

18. If an equipment grounding conductor is not run with a 4-wire, 3-phase feeder to a second building, can the grounded conductor of the feeder circuit be grounded to the same contiguous metal water pipe as the main electrical service in the first building where that water line runs to the second building?

Answer: _____

_____

_____

_____

Reference: _____

19. Many new dwellings have a conduit or tubing sleeve installed from the top of a panelboard up into the attic space so that future circuits can be gained from the panelboard without having to fish cables or cut sheetrock. Are such installations proper because future cables cannot be secured to the panelboard?

Answer: _____

_____

_____

_____

Reference: _____

20. Are all 15- and 20-ampere receptacles for dwelling-unit unfinished accessory buildings required to be GFCI protected?

    Answer: _____

    _____

    _____

    _____

    Reference: _____

2005 *CODE* CHANGES QUIZ 11

1. Where multiple-conductor power cables are installed in parallel, is it necessary for the equipment grounding conductor in each cable to be sized per the feeder or branch-circuit overcurrent device protecting the circuit according to Table 250.122?

    Answer: _____

    _____

    _____

    _____

    Reference: _____

2. What is a power panelboard?

    Answer: _____

    _____

    _____

    _____

    Reference: _____

3. Does the *Code* allow splicing of service-entrance conductors?

    Answer: _____

    _____

    _____

    _____

    Reference: _____

4. Is it permissible to use NM cable with temporary wiring on a construction site?

Answer: _____

_____

_____

_____

Reference: _____

5. Is it permissible to prewire electrical nonmetallic tubing before installing it in a concrete slab?

Answer: _____

_____

_____

_____

Reference: _____

6. Is there a limit on how many individual branch-circuits are permitted to be run from one building to another on the same property under single management?

Answer: _____

_____

_____

_____

Reference: _____

7. Why is MC cable allowed in assembly-type occupancies with over 100 occupants whereas Type AC cable is not allowed?

Answer: _____

_____

_____

_____

Reference: _____

8. Do the requirements that state *receptacle outlets shall not be installed in a face-up position in the work surfaces or countertops* apply to wet bar sink locations?

Answer: _____

_____

_____

_____

Reference: _____

9. What is the definition of a supervised industrial installation as opposed to a commercial establishment?

   Answer: _____

   _____

   _____

   _____

   Reference: _____

10. All 15- and 20-ampere, 125-volt receptacles in pediatric wards, rooms, or areas must be _____ or employ _____ covers.

    Answer: _____

    _____

    _____

    _____

    Reference: _____

11. Is it ever permissible to mount a small and lightweight luminaire (lighting fixture) to a device box, especially if the device box is a flush-mounted-type existing in a wall?

    Answer: _____

    _____

    _____

    _____

    Reference: _____

12. When trimming-in a job, how many inches (mm) of individual conductor must be left from the front of a 2 × 4 inch (50 × 100 mm) outlet box for the connection of or future replacement of toggle switches and receptacles?

    Answer: _____

    _____

    _____

    _____

    Reference: _____

13. Does the *NEC*® permit NM cable to be used for lease space wiring in a four-story building with 6000 square feet per floor that houses novelty shops?

    Answer: _____

    _____

    _____

    _____

    Reference: _____

14. Where in the *Code* is a "curb" around the floor opening required for busways passing vertically through more than one floor of all buildings?

Answer: _____

_____

_____

_____

Reference: _____

15. How many ground rods are required when the rod is to supplement a metallic water line electrode?

Answer: _____

_____

_____

_____

Reference: _____

16. Does Article 830—Network-Powered Broadband Communications Systems cover high-power communication systems?

Answer: _____

_____

_____

_____

Reference: _____

17. Is at least one general-care patient bed location receptacle outlet always required to be served by the normal system power supply, or can all general-care patient bed receptacles be served through emergency system connected panelboards?

Answer: _____

_____

_____

_____

Reference: _____

18. How many "main power feeders" may a single-family dwelling have and do the loads they serve make any difference? For example, a large single-family dwelling has a 600-ampere-rated service with two 100-ampere fused service switches, one 200-ampere fused switch, and a 400-ampere fused switch each feeding interior located panelboards. Are all "main power feeders"?

    Answer: _____

    _____

    _____

    _____

    Reference: _____

19. Is GFCI protection for 120-volt receptacle outlets required on the third floor outdoor balcony of a dwelling building?

    Answer: _____

    _____

    _____

    _____

    Reference: _____

20. Where in the *Code* are handholes covered?

    Answer: _____

    _____

    _____

    _____

    Reference: _____

PRACTICE EXAM 1

> Each lesson is designed purposely to require the student to apply the entire *NEC®* text, and not specific chapters or articles. It has been found that when studying for a timed, open-book examination, the student must gain proficiency in the Table of Contents, the Index, and the ability to move quickly from cover to cover to find the correct answer to each question in a timely fashion.

1. Are all 125-volt, 15- and 20-ampere receptacles in the service area of a commercial garage required to be protected by a GFCI?

   Answer: _____

   _____

   _____

   _____

   Reference: _____

2. A luminaire (lighting fixture) is installed over a hydromassage bathtub. Is GFCI protection required for the luminaire (fixture)?

   Answer: _____

   _____

   _____

   _____

   Reference: _____

3. A nongrounding-type receptacle is to be replaced with a GFCI receptacle because a grounding means does not exist within the box. Can you supply downstream receptacles from the GFCI? Are these downstream receptacles required to be 2-wire, nongrounding type?

   Answer: _____

   _____

   _____

   _____

   Reference: _____

4. A kitchen range has a built-in 125-volt receptacle and is within 6 feet (1.8 m) of the sink. Does the *Code* require such a receptacle to be protected with a GFCI?

   Answer: _____

   _____

   _____

   _____

   Reference: _____

5. Does the *NEC®* prohibit the use of nonmetallic outlet boxes with metal raceways?

Answer: _____

_____

_____

_____

Reference: _____

6. Does the *NEC®* permit luminaires (lighting fixtures) to be installed on trees?

Answer: _____

_____

_____

_____

Reference: _____

7. Electrical nonmetallic tubing (ENT) can be used concealed within walls, ceilings, and floors where the walls, ceilings, and floors provide a thermal barrier of material that has at least _____.

Answer: _____

_____

_____

_____

Reference: _____

8. Does the *Code* require any extra protection for electrical nonmetallic tubing (ENT) when it is installed through openings in metal studs?

Answer: _____

_____

_____

_____

Reference: _____

9. What are the strapping requirements for ENT when installed in metal studs?

Answer: _____

_____

_____

_____

Reference: _____

10. Does the *Code* require any extra protection for Type NM cable when it is installed through openings in metal studs?

Answer: _____

_____

_____

_____

Reference: _____

11. When UF cable is used for interior wiring, are the conductors required to be rated at 90°C?

Answer: _____

_____

_____

_____

Reference: _____

12. Must "hospital grade" receptacles be installed throughout hospitals?

Answer: _____

_____

_____

_____

Reference: _____

13. When railings or lattice work is used for room dividers, does the *Code* require receptacles to be installed as if they were solid walls?

Answer: _____

_____

_____

_____

Reference: _____

14. In an apartment project, multiwire branch-circuits are routed through outlet boxes to supply receptacles. Can the neutral conductor continuity be assured by two connections to the screw terminals of the receptacle?

Answer: _____

_____

_____

_____

Reference: _____

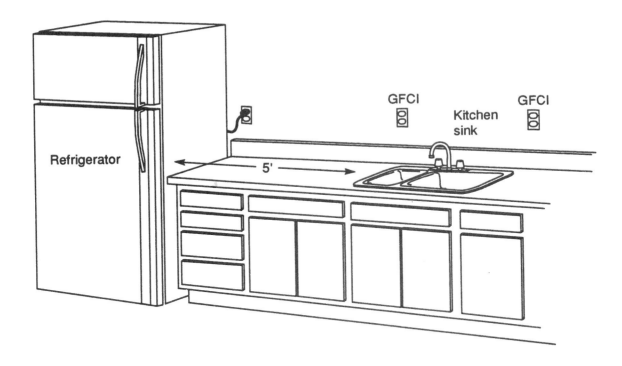

15. As shown in the diagram above, a small appliance circuit receptacle is above the counter next to the refrigerator. The refrigerator is plugged into this receptacle that is 5 feet (1.5 m) from the kitchen sink. Is the receptacle required to have GFCI protection?

Answer: _____
_____
_____
_____

Reference: _____

16. Does the *Code* require outdoor receptacles to be installed on balconies of high-rise apartment buildings?

Answer: _____
_____
_____
_____

Reference: _____

17. The *Code* did require that switches and circuit breakers used as switches be installed so that the center of the grip of the operating handle when in its highest position was no more than 6 feet (1.8 m) above the floor or working platform. Is this still the case?

Answer: _____
_____
_____
_____

Reference: _____

18. Where can luminaires (lighting fixtures) be used as raceways for circuit conductors?

Answer: _____

_____

_____

_____

Reference: _____

19. In what areas of a hospital are hospital-grade receptacles required?

Answer: _____

_____

_____

_____

Reference: _____

20. Is it a requirement that receptacles over a wet bar be spaced according to cabinet or floor line requirements?

Answer: _____

_____

_____

_____

Reference: _____

21. A weatherproof receptacle cover must be weatherproof when in use (attachment cap inserted) or shall have a _____ feature.

Answer: _____

_____

_____

_____

Reference: _____

22. Would electrical nonmetallic tubing be an acceptable wiring method for a wet-niche luminaire (lighting fixture) or some portion of the circuit?

Answer: _____

_____

_____

_____

Reference: _____

23. Can nonmetallic-sheathed cable be used to supply a recessed fluorescent luminaire (fixture)?

Answer: _____

_____

_____

_____

Reference: _____

24. Do conductors for festoon lighting have to be rubber covered?

Answer: _____

_____

_____

_____

Reference: _____

25. Can an electric discharge luminaire (fixture) such as a fluorescent strip be used as a raceway for circuit conductors?

Answer: _____

_____

_____

_____

Reference: _____

PRACTICE EXAM 2

1. Are all luminaires (fixtures) located in agriculture buildings to be listed as dusttight or watertight?

Answer: _____

_____

_____

_____

Reference: _____

2. When installing AC cable horizontally through metal studs, are insulators needed?

Answer: _____

_____

_____

_____

Reference: _____

3. What types of cable assemblies are acceptable for use as feeders to a floating building?

Answer: _____

_____

_____

_____

Reference: _____

4. Would nonmetallic-sheathed cable be acceptable to wire a pool-associated motor?

Answer: _____

_____

_____

_____

Reference: _____

5. Would liquidtight flexible conduit be acceptable for enclosing conductors from an adjacent weatherproof box to an approved swimming pool junction box?

Answer: _____

_____

_____

_____

Reference: _____

6. Is the cord-and-plug connection of an electric water heater acceptable under the *NEC*®? If so, would a standard dryer cord be acceptable (240 volts with neutral)? (Assume that ampacities, and so on, are acceptable under the *NEC*®.)

Answer: _____

_____

_____

_____

Reference: _____

7. Does the *Code* permit motor controllers and disconnects to be used as junction boxes or wireways?

Answer: _____

_____

_____

_____

Reference: _____

8. In determining the locked-rotor kVA per horsepower of a dual voltage electric motor, which voltage must be considered?

Answer: _____

_____

_____

_____

Reference: _____

9. Would electrical nonmetallic tubing be acceptable to enclose the insulated conductors for a panelboard (not part of service equipment) feeding pool-associated equipment?

Answer: _____

_____

_____

_____

Reference: _____

10. Does the *Code* permit the installation of 15-ampere-rated receptacles on the 20-ampere small appliance branch-circuits required in a dwelling?

Answer: _____

_____

_____

_____

Reference: _____

11. Does the *Code* permit the installation of a 15-ampere-rated duplex receptacle on a 20-ampere separate circuit for a microwave oven with a nameplate reading of 13 amperes?

Answer: _____

_____

_____

_____

Reference: _____

12. What does the term "in free air" mean at the top of Table 310.17?

Answer: _____

_____

_____

_____

Reference: _____

13. Has the *National Electrical Code*® restricted rigid nonmetallic conduit (PVC) from being installed in cold weather areas?

Answer: _____

_____

_____

_____

Reference: _____

14. All temporary wiring 15-ampere and 20-ampere, 125-volt receptacles not a part of the permanent building structure need to have GFCI protection. Does this mean that if a permanently installed receptacle is installed and energized, no GFCI protection is required even though this outlet is to be used by construction personnel?

Answer: _____

_____

_____

_____

Reference: _____

15. In the *NEC*®, would a GFCI-protected receptacle be required on a wet bar in a dwelling?

Answer: _____

_____

_____

_____

Reference: _____

16. Must the garbage disposal receptacle located under the sink be GFCI protected because it is within 6 feet (1.8 m) of the sink?

Answer: _____

_____

_____

_____

Reference: _____

17. What is the maximum number of duplex receptacles permitted on a 2-wire, 15-ampere circuit in a dwelling?

Answer: _____

_____

_____

_____

Reference: _____

18. How many receptacles are required at the patient's bed location in a critical care facility?

Answer: _____

_____

_____

_____

Reference: _____

19. A ¼-horsepower circulating pump is to be located in a closet space housing a gas-fired water heater in a large single-family residence. Can the circulating pump be cord-and-plug connected?

Answer: _____

_____

_____

_____

Reference: _____

20. In lieu of switching a ceiling luminaire (lighting fixture), the *Code* permits switching a receptacle. Can the switched receptacle be counted as one of the receptacles required for the 12-foot (3.6 m) rule?

Answer: _____

_____

_____

_____

Reference: _____

21. A laundry room in a single-family residence has a receptacle outlet on one wall and a second receptacle on the opposite wall. Can these two outlets be fed by the same laundry branch outlet?

Answer: _____

_____

_____

_____

Reference: _____

22. Antique stores and others are selling hanging luminaires (fixtures) without a grounding conductor. Are these luminaires (fixtures) in violation of the *NEC*®?

Answer: _____

_____

_____

_____

Reference: _____

23. Can 6 feet (1.8 m) of flexible metal conduit be installed between rigid nonmetallic raceways used for service-entrance conductors?

Answer: _____

_____

_____

_____

Reference: _____

24. Many kitchen and bathroom sinks, tubs, and showers are provided with short sections of metal pipe (6 inches [150 mm] to 2 feet [600 mm]) for connection to nonmetallic pipe systems. Is it required to bond the several short sections of metal pipe to the service equipment?

Answer: _____

_____

_____

_____

Reference: _____

25. Is there a limit on the number of extension boxes that can be joined together?

Answer: _____

_____

_____

_____

Reference: _____

PRACTICE EXAM 3

1. A switch box was installed in a floor of a dwelling. A receptacle outlet with a standard plate was installed on the box. Does this standard outlet box and receptacle installed in the floor comply with the *Code*?

Answer: _____

_____

_____

_____

Reference: _____

2. A 4 inch × 2⅛ inch-square NM box is marked with the number of 14, 12, and 10 AWG conductors that can be installed in it. Does this mean that 8 AWG conductors are not permitted in the box?

Answer: _____

_____

_____

_____

Reference: _____

3. Are outdoor receptacle outlets required at each dwelling unit of a multifamily dwelling?

Answer: _____

_____

_____

_____

Reference: _____

4. Do portable lamps wired with flexible cord require a polarized attachment plug?

Answer: _____

_____

_____

_____

Reference: _____

5. When figuring receptacles for a bedroom in a residence, must you include the space behind the door?

Answer: _____

_____

_____

_____

Reference: _____

6. Are metal boxes required for splices in temporary wiring?

Answer: _____

_____

_____

_____

Reference: _____

7. What is the maximum number of 12 AWG conductors permitted in a 4 inch × 1½ inch-deep octagon box?

Answer: _____

_____

_____

_____

Reference: _____

8. When a junction box contains a combination of conductors, such as 14 AWG and 12 AWG wires, what table is used to determine the size of the box?

Answer: _____

_____

_____

_____

Reference: _____

9. Can you use threaded intermediate metal conduit in a Class 1, Division 1 location?

Answer: _____

_____

_____

_____

Reference: _____

10. Does the *Code* consider EMT as conduit?

Answer: _____

_____

_____

_____

Reference: _____

11. Can suspended fluorescent luminaires (fixtures) be connected together with unsupported EMT between them?

Answer: _____

_____

_____

_____

Reference: _____

12. Must lay-in fluorescent luminaires (lighting fixtures) be fastened to the grid T-bars of a suspended ceiling system?

Answer: _____

_____

_____

_____

Reference: _____

13. Can luminaires (fixtures) marked "suitable for damp locations" be used in wet locations?

Answer: _____

_____

_____

_____

Reference: _____

14. A UL-listed fluorescent luminaire (fixture) is equipped with a Class "P" ballast. Does the *Code* permit such a luminaire (fixture) to be surface mounted on a combustible low-density cellulose fiberboard ceiling?

Answer: _____

_____

_____

_____

Reference: _____

15. Are all recessed incandescent luminaires (fixtures) required to have thermal protection?

Answer: _____

_____

_____

_____

Reference: _____

16. Is it necessary to identify the high leg of a 240/120-volt system at the motor disconnect of a 3-phase motor?

Answer: _____

_____

_____

_____

Reference: _____

17. Is a GFCI protection required for a receptacle located on a food preparation island that is less than 6 feet (1.8 m) from the countertop sink?

Answer: _____

_____

_____

_____

Reference: _____

18. Can conductors that pass through a panelboard also be spliced in the panelboard?

Answer: _____

_____

_____

_____

Reference: _____

19. Are the enclosures for mercury vapor luminaires (lighting fixtures) installed 8 feet (2.4 m) above grade on poles outdoors required to be grounded?

Answer: _____

_____

_____

_____

Reference: _____

20. Would an exit luminaire (light) with an emergency pack capable of supplying more than 1½ hours of power have to be connected to an emergency circuit ahead of the main disconnect?

Answer: _____

_____

_____

_____

Reference: _____

21. A grounded 20-ampere receptacle is installed in a residential garage to serve an air compressor. Would this receptacle require GFCI protection?

Answer: _____

_____

_____

_____

Reference: _____

22. If additional receptacles are installed in a bathroom, not adjacent to the wash basin, are they required to be GFCI protected?

    Answer: _____

    _____

    _____

    _____

    Reference: _____

23. Must a receptacle in a dwelling-unit kitchen installed for a luminaire (lighting fixture) be GFCI protected?

    Answer: _____

    _____

    _____

    _____

    Reference: _____

24. When installing recessed luminaires (fixtures) (hi-hats) in the ceiling, does 314.20 apply to the luminaire (lighting fixture) housing?

    Answer: _____

    _____

    _____

    _____

    Reference: _____

25. What are the support requirements for service-entrance cables?

    Answer: _____

    _____

    _____

    _____

    Reference: _____

PRACTICE EXAM 4

1. An underground cable is routed from the service equipment in the building underground and up a pole in a commercial parking lot to supply an overhead luminaire (light). What is the minimum burial depth for the cable below the base of the pole?

   Answer: _____

   _____

   _____

   _____

   Reference: _____

2. Does the *Code* permit 15-ampere duplex receptacles to be installed on a 20-ampere branch-circuit?

   Answer: _____

   _____

   _____

   _____

   Reference: _____

3. We are designing a new hotel with 250 guest rooms. It is our intent to install permanent beds, desks, and dressers in each guest room, bolted to the wall. If receptacles are located in accordance with 210.52, the receptacles will be located behind the headboard of the bed and behind the dresser. Are the receptacles required behind these pieces of furniture? Would the number of required receptacles be less if the furniture was not bolted to the wall?

   Answer: _____

   _____

   _____

   _____

   Reference: _____

4. Can a paddle fan be supported only by an outlet box if it weighs less than 35 pounds (16 kg)?

   Answer: _____

   _____

   _____

   _____

   Reference: _____

5. Where can luminaires (lighting fixtures) be used as raceways for circuit conductors?

   Answer: _____

   _____

   _____

   _____

   Reference: _____

6. Do recessed incandescent luminaires (lighting fixtures) require thermal protection?

   Answer: _____

   _____

   _____

   _____

   Reference: _____

7. Are medium-base lampholders in industrial occupancies permitted on 277-volt luminaire (lighting) circuits?

Answer: _____

_____

_____

_____

Reference: _____

8. Can a battery-powered (unit equipment) emergency light be supplied directly from a circuit in a panelboard that supplies normal lumination (lighting) in the area that will be illuminated by the battery luminaire (light) under emergency conditions?

Answer: _____

_____

_____

_____

Reference: _____

9. A single transfer switch is permitted to serve one or more branches of the essential electrical system in a small hospital. What constitutes a small hospital?

Answer: _____

_____

_____

_____

Reference: _____

10. Would a 3-phase, 4-wire, 1200-ampere, 480/277-volt service switch with 900-ampere fuses be required to have ground-fault protection?

Answer: _____

_____

_____

_____

Reference: _____

11. Are luminaires (lighting fixtures) operating at less than 15 volts between conductors required to be protected by a GFCI over hot tubs or spas?

Answer: _____

_____

_____

_____

Reference: _____

12. Could you use electrical metallic tubing (EMT) in the earth below a concrete slab?

    Answer: _____

    _____

    _____

    _____

    Reference: _____

13. An industrial building will be served by several services of different voltage ratings. Is it required that each system be separately grounded?

    Answer: _____

    _____

    _____

    _____

    Reference: _____

14. In residential occupancies, locating equipment to comply with the *Code* becomes a problem. Can a service panel or distribution panelboard be located over a counter or an appliance (washing machine, dryer) that extends away from the wall?

    Answer: _____

    _____

    _____

    _____

    Reference: _____

15. A surface-mounted fluorescent luminaire (fixture) with an integral receptacle is installed above the countertop in the kitchen of a dwelling and is not more than 5½ feet (1.7 m) above the floor. Is this luminaire (fixture) required to be supplied by a 20-ampere circuit as are other kitchen receptacles? Is the luminaire (fixture) receptacle required to be protected by GFCI?

    Answer: _____

    _____

    _____

    _____

    Reference: _____

16. What are the requirements for grounding an agricultural building where livestock are housed?

    Answer: _____

    _____

    _____

    _____

    Reference: _____

17. Can a fused disconnect switch or circuit breaker located on the meter pole of a farm installation be used as the service equipment for the residence?

    Answer: _____

    _____

    _____

    _____

    Reference: _____

18. A Type UF cable feeds a 120-volt yard luminaire (light) location on residential property. The circuit is protected by a 15-ampere overcurrent protection device. It is not GFCI protected. What is the minimum burial depth permitted for the cable?

    Answer: _____

    _____

    _____

    _____

    Reference: _____

19. Can a fixed storage-type water heater having a capacity of 120 gallons or less be cord-and-plug connected?

    Answer: _____

    _____

    _____

    _____

    Reference: _____

20. A TV dish antenna is located in the side yard of a single-family dwelling. (a) What is the minimum depth required by the *NEC®* for the coaxial signal cable? (b) How are these antenna units required to be grounded?

    Answer: _____

    _____

    _____

    _____

    Reference: _____

21. Does the *Code* permit low-voltage control cables to be supported from the conduit containing the power circuit conductors feeding an air-conditioning unit?

    Answer: _____

    _____

    _____

    _____

    Reference: _____

22. Is it true that if the underground metallic water piping is not at least 10 feet (3 m) long, the underground piping system is not adequate as a grounding electrode?

Answer: _____

_____

_____

_____

Reference: _____

23. Does the luminaire (lighting) outlet in the attic have to be switch controlled?

Answer: _____

_____

_____

_____

Reference: _____

24. With regards to recessed luminaires (lighting fixtures) installed over showers, is it permissible to use a gasketed vapor-proof trim only to comply with the *Code* or must the entire luminaire (fixture) be rated for a wet location?

Answer: _____

_____

_____

_____

Reference: _____

25. May overhead outside branch-circuit conductors of 10 AWG copper be used in an unsupported open span up to 50 feet (15 m)? Is 10 AWG listed copper multiconductor UF cable suitable for such an application?

Answer: _____

_____

_____

_____

Reference: _____

PRACTICE EXAM 5

1. Is a nonmetallic cable tray permitted by the *NEC*®?

Answer: _____

_____

_____

_____

Reference: _____

2. The *Code* requires clearance for luminaires (lighting fixtures) installed in clothes closets. (a) What are the clearance requirements for surface-mounted luminaire (fixtures) mounted on the wall above the door? (b) What are clearance requirements for surface-mounted luminaire (fixtures) mounted on the ceiling? (c) What are the clearance requirements for recessed incandescent or fluorescent luminaire (fixtures) ceiling mounted?

Answer: _____

_____

_____

_____

Reference: _____

3. Are 16 AWG luminaire (fixture) wires counted for the number of conductors in outlet boxes to determine the box fill?

Answer: _____

_____

_____

_____

Reference: _____

4. Are switchboards and control panels rated 1200 amperes or more, 600 volts, and over 6 feet (1.8 m) wide required to have one entrance at each end of the room?

Answer: _____

_____

_____

_____

Reference: _____

5. What size grounding-electrode conductor is required where a grounding conductor is routed to a driven rod and from the rod to a concrete-encased electrode?

Answer: _____

_____

_____

_____

Reference: _____

6. Can handle locks on circuit breakers be locked so that the power to loads such as emergency lighting, sump pumps, alarm warning circuits, and other types of equipment cannot be cut off by mistake? What section of the *NEC*® applies?

Answer: _____

_____

_____

_____

Reference: _____

7. Can a service be mounted on a mobile home if the home is designed to be installed on a foundation?

Answer: _____

_____

_____

_____

Reference: _____

8. Is an insulated equipment ground required for a pool panelboard feeder fed from the service equipment?

Answer: _____

_____

_____

_____

Reference: _____

9. Would the *Code* allow the use of stainless steel ground rods?

Answer: _____

_____

_____

_____

Reference: _____

10. The *Code* now allows flexible metal conduit for services. Does the *Code* permit liquidtight flexible conduit, metallic or nonmetallic?

Answer: _____

_____

_____

_____

Reference: _____

11. The grounding conductor can be connected to the grounding electrode by listed connectors, clamps, or other listed means and _____.

    Answer: _____

    _____

    _____

    _____

    Reference: _____

12. Does the *Code* require a ground wire in all flexible raceways?

    Answer: _____

    _____

    _____

    _____

    Reference: _____

13. Is AC cable a permitted wiring method in places of assembly?

    Answer: _____

    _____

    _____

    _____

    Reference: _____

14. Does 240.4(D) apply to motor circuits?

    Answer: _____

    _____

    _____

    _____

    Reference: _____

15. Can liquidtight flexible nonmetallic conduit be used on residential work?

    Answer: _____

    _____

    _____

    _____

    Reference: _____

16. Are there any *Code* requirements against installing a bare neutral wire inside the service conduit?

    Answer: _____

    _____

    _____

    _____

    Reference: _____

17. Can the branch-circuit and Class II control wires for two 3-phase motors be installed in a common raceway of the proper size?

    Answer: _____

    _____

    _____

    _____

    Reference: _____

18. The *Code* explicitly requires bonding around concentric knockouts on service equipment, but does the *Code* require that equipment downstream from the service equipment be bonded in the same manner?

    Answer: _____

    _____

    _____

    _____

    Reference: _____

19. Is nonmetallic-sheathed cable permitted to run through cold air returns if it is sleeved with thin-wall conduit or greenfield in short lengths?

    Answer: _____

    _____

    _____

    _____

    Reference: _____

20. Is it necessary to pigtail the neutral when connecting receptacles to a multiwire circuit?

    Answer: _____

    _____

    _____

    _____

    Reference: _____

21. When a motor-control circuit extends beyond the control enclosure, what is the maximum overcurrent protection acceptable for 12 AWG copper control wires?

    Answer: _____

    _____

    _____

    _____

    Reference: _____

22. Is an equipment bonding conductor required to bond a cover-mounted receptacle to a surface-mounted outlet box?

    Answer: _____

    _____

    _____

    _____

    Reference: _____

23. When calculating branch-circuit and feeder loads, what are the normal system voltages that shall be used?

    Answer: _____

    _____

    _____

    _____

    Reference: _____

24. Is it permissible to run four 12 AWG THW copper conductors in a cable or raceway where they encounter 100°F ambient temperatures to supply cord-and-plug-connected loads protected at 20 amperes?

    Answer: _____

    _____

    _____

    _____

    Reference: _____

25. When flat conductor cable, Type FCC, is installed under carpet, what is the maximum size carpet squares permitted?

    Answer: _____

    _____

    _____

    _____

    Reference: _____

PRACTICE EXAM 6

1. Are bonding jumpers required on metal feeder and branch-circuit raceways containing circuits of more than 250 volts-to-ground where oversize concentric or eccentric knockouts are encountered?

   Answer: _____

   _____

   _____

   _____

   Reference: _____

2. What is the maximum number of service disconnects allowed for a set of service-entrance conductors installed in an apartment per the *NEC*®?

   Answer: _____

   _____

   _____

   _____

   Reference: _____

3. (a) Is a GFCI required on or within an outdoor portable sign? (b) Would a GFCI-protected branch-circuit supplying the sign be acceptable?

   Answer: _____

   _____

   _____

   _____

   Reference: _____

4. Is a luminaire (lighting) outlet required in the crawl space of a manufactured building if equipment requiring service is installed in the space?

   Answer: _____

   _____

   _____

   _____

   Reference: _____

5. Can an office partition assembly be cord-and-plug connected?

   Answer: _____

   _____

   _____

   _____

   Reference: _____

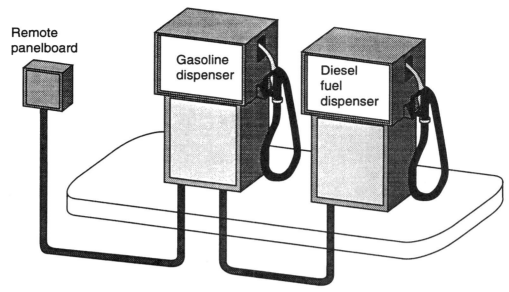

6. In the diagram above, a diesel fuel dispenser is mounted adjacent to a gasoline dispenser on a service station island. A sealing fitting is installed in the conduit entering the gasoline dispenser. Is a sealing fitting required in the conduit entering the diesel fuel dispenser?

Answer: _____

_____

_____

_____

Reference: _____

7. In residential-type occupancies, does the wall space behind doors where the door opens against the wall count when spacing the receptacles in a room?

Answer: _____

_____

_____

_____

Reference: _____

8. A four-family apartment building is supplied by an overhead drop to a 300-kcmil THW aluminum riser. Four electric meters are grouped and have 100-ampere breakers below the meters to supply a panel in each apartment. 3 AWG THW copper feeders are run to each apartment unit. An individual water meter is provided for each apartment. Each water meter is supplied with nonmetallic pipe. The water system for each apartment is copper above grade. (a) What size grounding electrode conductor is required? (b) How would this conductor be attached to the metallic water piping in each apartment? (c) Is a supplementary driven electrode required?

Answer: _____

_____

_____

_____

Reference: _____

9. Is a 14 AWG copper conductor allowed in a kitchen of a dwelling unit if it serves a single fixed appliance such as a dishwasher?

Answer: _____

_____

_____

_____

Reference: _____

10. Can Type UF cable be used to supply a swimming pool motor?

Answer: _____

_____

_____

_____

Reference: _____

11. Are 16 AWG and 18 AWG luminaire (lighting fixture) wires counted for the number of conductors in outlet boxes for box fill?

Answer: _____

_____

_____

_____

Reference: _____

12. Is a receptacle outlet required for a wet bar and/or if an outlet is installed within 6 feet (1.8 m) of the wet bar sink, would it be required to be GFCI protected?

Answer: _____

_____

_____

_____

Reference: _____

13. When calculating the demand for a 3-kW and a 6-kW range in a dwelling, would you use Table 220.55, Column B for the 3-kW and Column C for the 6-kW range or would you combine them and use either B or C for the two ranges?

Answer: _____

_____

_____

_____

Reference: _____

14. When sizing the branch-circuit conductors to a 4.5-kW dryer, would you use 4.5 kW or 5 kW in making the calculations?

Answer: _____

_____

_____

_____

Reference: _____

15. What is the maximum weight of a ceiling fan that can be supported by an outlet box?

Answer: _____

_____

_____

_____

Reference: _____

16. Where does the *NEC*® list or calculate the number of conductors allowed in flexible metal conduit?

Answer: _____

_____

_____

_____

Reference: _____

17. Are ceiling grid support wires that rigidly support a lay-in ceiling acceptable for the sole support of junction boxes above the lay-in ceiling?

Answer: _____

_____

_____

_____

Reference: _____

18. Can electrical equipment approved for a Class I location be used in a Class II location?

Answer: _____

_____

_____

_____

Reference: _____

19. Can THHN insulated conductors be used outside for service-entrance conductors where exposed to the elements?

    Answer: _____

    _____

    _____

    _____

    Reference: _____

20. Who has the authority to determine acceptability of electrical equipment and materials?

    Answer: _____

    _____

    _____

    _____

    Reference: _____

21. Can more than one receptacle in a laundry room be supplied by the laundry branch-circuit?

    Answer: _____

    _____

    _____

    _____

    Reference: _____

22. Many air-conditioning units are being installed on roof tops of apartment buildings for different reasons. Is a receptacle adjacent to such equipment required for servicing this equipment?

    Answer: _____

    _____

    _____

    _____

    Reference: _____

23. *NEC®* 430.72 applies to motor-control circuits that are tapped from the load side of motor branch-circuit short-circuit, and ground-fault protective devices. Because Column B of Table 430.72(B) covers control circuits that do not extend beyond the motor control equipment enclosure and Column C covers those that do extend beyond, what does Column A cover?

    Answer: _____

    _____

    _____

    _____

    Reference: _____

24. A 240-volt, single-phase circuit with a rated load of 1.8 kW draws 9 amperes of current. The power factor is _____ percent.

Answer: _____

_____

_____

_____

Reference: _____

25. Can you run multioutlet assembly from one room to another room through a sheetrock wall?

Answer: _____

_____

_____

_____

Reference: _____

PRACTICE EXAM 7

1. Illuminated·exit signs are required by local building codes, and battery packs will be used. Do these signs have to be installed on a luminaire (lighting) circuit in the area of the sign or could they be on a separate branch-circuit?

Answer: _____

_____

_____

_____

Reference: _____

2. Can a metal cable tray be used as an equipment grounding conductor?

Answer: _____

_____

_____

_____

Reference: _____

3. Can surface nonmetallic raceway be used outside in a wet location?

Answer: _____

_____

_____

_____

Reference: _____

4. Switchboards and control panels rated 1200 amperes or more, 600 volts or less, and over 6 feet (1.8 m) wide are required to have one entrance at each end of the room. What conditions would permit only one entrance to this room?

Answer: _____

_____

_____

_____

Reference: _____

5. Is Type MC cable required to have a bushing installed like the bushings used on Type AC armored cable?

Answer: _____

_____

_____

_____

Reference: _____

6. Can electrical nonmetallic tubing be surface mounted in a warehouse building?

Answer: _____

_____

_____

_____

Reference: _____

7. How deep must a residential branch-circuit be buried if it passes under the driveway?

Answer: _____

_____

_____

_____

Reference: _____

8. Where more than one building or structure is on the same property and under single management, the *Code* requires each building to have a disconnecting means at each such building or structure of 600 volts or less. Under what conditions can the disconnecting means be located elsewhere on the premises?

Answer: _____

_____

_____

_____

Reference: _____

9. Are cable trays permitted in a grain elevator?

   Answer: _____

   _____

   _____

   _____

   Reference: _____

10. If a luminaire (fixture) is listed for use as a raceway, can this raceway be used for conductors to supply a circuit or circuits beyond the luminaire (fixture)?

    Answer: _____

    _____

    _____

    _____

    Reference: _____

11. What *Code* section prohibits bare neutral feeders?

    Answer: _____

    _____

    _____

    _____

    Reference: _____

12. Would a large mercantile store be considered a place of assembly if it can hold more than 100 people?

    Answer: _____

    _____

    _____

    _____

    Reference: _____

13. Can the neutral be reduced by 70 percent for data processing equipment over 200 amperes?

    Answer: _____

    _____

    _____

    _____

    Reference: _____

14. What ampacity correction factor would you use for 12 AWG THW copper conductors used in an ambient temperature of 78°F? (Not more than three conductors in raceway, in free air, 240-volt circuit.)

    Answer: _____

    _____

    _____

    _____

    Reference: _____

15. How are box volumes calculated when the box contains different size conductors and a combination of clamps, devices, and studs?

    Answer: _____

    _____

    _____

    _____

    Reference: _____

16. What is the minimum service-drop clearance required over a residential yard where the drop conductors do not exceed 150 volts-to-ground?

    Answer: _____

    _____

    _____

    _____

    Reference: _____

17. An equipment room containing electrical, telephone, and air-handling equipment uses the space in the room for air-handling purposes. Is this room considered a plenum?

    Answer: _____

    _____

    _____

    _____

    Reference: _____

18. As applied to lumination (lighting) in a clothes closet, is the entire 12-inch (300 mm) or width-of-shelf area required to be unobstructed to the floor or could part of this area be used for storage?

    Answer: _____

    _____

    _____

    _____

    Reference: _____

19. Would an outdoor luminaire (lighting fixture) be considered as grounded by the physical connection to a grounded metal support pole or would an equipment grounding conductor connection directly to the luminaire (fixture) be required?

Answer: _____

_____

_____

_____

Reference: _____

20. Would entrance-type luminaires (lighting fixtures) that are installed under roof overhangs, porch roofs, or canopies be required to be suitable for damp locations?

Answer: _____

_____

_____

_____

Reference: _____

21. Table 430.147 through Table 430.150 list the horsepower and the full-load currents of motors. How could you determine the full-load current of a motor not listed, such as a 35-horsepower, 460-volt motor?

Answer: _____

_____

_____

_____

Reference: _____

22. Where a single Type NM cable is pulled into a raceway, what percent may the cross section of the conduit be filled?

Answer: _____

_____

_____

_____

Reference: _____

23. Can you use Table 310.15(B)(6) for a feeder to each apartment in a building regardless of whether the mains are in the apartment units or on the outside of the building?

Answer: _____

_____

_____

_____

Reference: _____

24. Can swimming pool equipment such as a pump motor be calculated as a fixed appliance when applying 220.14(A) to a dwelling?

   Answer: _____

   _____

   _____

   _____

   Reference: _____

25. Is a GFCI receptacle required in a detached garage of a dwelling?

   Answer: _____

   _____

   _____

   _____

   Reference: _____

PRACTICE EXAM 8

1. Are all recessed luminaires (lighting fixtures) that are to be installed in a suspended ceiling required to be thermally protected?

   Answer: _____

   _____

   _____

   _____

   Reference: _____

2. How can you ground an agricultural building and comply with the *Code*?

   Answer: _____

   _____

   _____

   _____

   Reference: _____

3. Is Table 220.55 applicable to microwave ovens and convection cooking ovens?

   Answer: _____

   _____

   _____

   _____

   Reference: _____

4. Is an autotransformer recognized for use in a motor-control circuit?

Answer: _____

_____

_____

_____

Reference: _____

5. Can service equipment be mounted on a floating dwelling unit that is moved frequently?

Answer: _____

_____

_____

_____

Reference: _____

6. Is it permissible to use neon lumination (lighting) on or in a dwelling unit?

Answer: _____

_____

_____

_____

Reference: _____

7. In a large industrial plant, does the *Code* allow a branch-circuit over 50 amperes to feed several outlets such as to supply power to portable welders that move from place to place?

Answer: _____

_____

_____

_____

Reference: _____

8. Can nonmetallic raceways be used in healthcare facilities?

Answer: _____

_____

_____

_____

Reference: _____

9. Can a conductor be protected by an overcurrent device sized for its ampacity even though the allowable load current is only 50 percent of the assigned ampacity after derating?

Answer: _____

_____

_____

_____

Reference: _____

10. Does the *Code* permit single-pole circuit breakers in lighting panelboards to be used as a switch for controlling a lighting circuit as an off and on switch?

Answer: _____

_____

_____

_____

Reference: _____

11. When determining the required size box, items such as fittings, cable clamps, hickeys, where combinations of different size conductors are used, which conductor size do you deduct for the device and fittings?

Answer: _____

_____

_____

_____

Reference: _____

12. Where rigid metal conduit is used as service raceways, are double locknuts permitted for the service equipment required continuity?

Answer: _____

_____

_____

_____

Reference: _____

13. A receptacle outlet is to be added to an existing installation, and Type NM cable is to be "fished in" the partition wall. Does the *Code* allow a nonmetallic box without cable clamps to be used for such an installation?

Answer: _____

_____

_____

_____

Reference: _____

14. What size general-use switch is required on an air-conditioning compressor, sealed (hermetic-type) when the nameplate states FLA of 100 amperes, 3-phase, 230 volts, LRA 580?

Answer: _____

_____

_____

_____

Reference: _____

15. A fluorescent luminaire (fixture) is supplied with trade size $3/8$ (12) flexible metal conduit 5 feet (1.5 m) long. Is an equipment grounding required to ground the luminaire (fixture)?

Answer: _____

_____

_____

_____

Reference: _____

16. What is the full-load current of an electric furnace rated 240 volts, single-phase, 10 kW when connected to a 208-volt circuit? (Show calculations.)

Answer: _____

_____

_____

_____

Reference: _____

17. Can four 400-watt wet-niche luminaires (lights) be connected to one GFCI for a swimming pool?

Answer: _____

_____

_____

_____

Reference: _____

18. How does one determine how many conductors conduit bodies (condulets) are approved for?

Answer: _____

_____

_____

_____

Reference: _____

19. How can a recessed luminaire (lighting fixture) approved for direct contact with thermal insulation be identified?

Answer: _____

_____

_____

_____

Reference: _____

20. Does the 10-foot (3 m) tap rule apply only to conductors such as wires and not panelboard bus?

Answer: _____

_____

_____

_____

Reference: _____

21. How are working clearances measured between enclosed electrical equipment facing each other across an aisle?

Answer: _____

_____

_____

_____

Reference: _____

22. Can either the disposal, dishwasher, or trash compactor be installed on a 20-ampere kitchen small appliance circuit?

Answer: _____

_____

_____

_____

Reference: _____

23. What is the required horizontal distance between service drops and swimming pools?

Answer: _____

_____

_____

_____

Reference: _____

24. Can a bare 4 AWG copper grounding electrode conductor be buried below a concrete floor slab, and is there any depth requirement?

Answer: _____

_____

_____

_____

Reference: _____

25. Where the grounding electrode conductor connects to the water system on the house side of the water meter, is bonding required around valves with sweated connections? Also, what other equipment is required to be bonded around?

Answer: _____

_____

_____

_____

Reference: _____

PRACTICE EXAM 9

1. Can a floodlight and/or receptacle be on the GFCI circuit protecting an underwater luminaire (lighting fixture)?

Answer: _____

_____

_____

_____

Reference: _____

2. For a 600-ampere service where 350 kcmil copper conductors are parallel in two conduits and only 3-phase, 3-wire is needed, but the transformer is Y-connected and grounded: (a) Must the grounded conductor be brought into the service? (b) Can the grounded conductor be brought in through a single raceway or is it required to be installed in each of the parallel raceways? (c) What is the maximum size of the grounded conductor?

Answer: _____

_____

_____

_____

Reference: _____

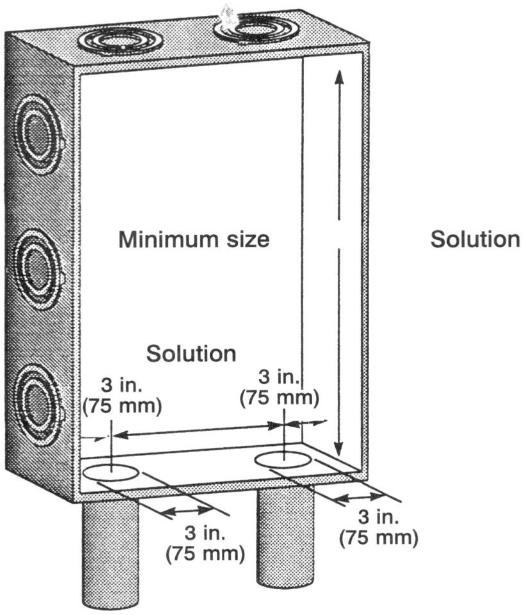

3. What are the minimum dimensions of a junction box illustrated above with only a U-pull of two trade size 3 (78) conduits in one side?

Answer: _____

_____

_____

_____

Reference: _____

4. With two 42-circuit panelboards bolted together, can the 42 circuits from one panelboard feed through the gutter of the second panelboard if the 40 percent fill in that gutter is not exceeded?

Answer: _____

_____

_____

_____

Reference: _____

5. Can suspended fluorescent luminaires (fixtures) be connected together with unsupported EMT between them if the length of the EMT is less than 3 feet (900 mm)?

Answer: _____

_____

_____

_____

Reference: _____

6. (a) Are 3-phase generators permitted to be connected to a three-pole transfer switch with a solid neutral or is a four-pole transfer switch that breaks the neutral required? (b) If either is OK, must a grounding electrode be provided at the generator in both cases?

Answer: _____

_____

_____

_____

Reference: _____

7. A 600-ampere underground service with four 4/0 AWG conductors per phase paralleled is installed in two rigid nonmetallic conduits. Must the phase conductors and neutral be grouped in a single raceway if the conduits are nonmetallic?

Answer: _____

_____

_____

_____

Reference: _____

8. The raceway for a 240-volt motor circuit includes both the power circuit and control circuit conductors. The power circuit conductors use 600-volt insulation. Must the control circuit conductors also have 600-volt insulation or would 300-volt insulation be acceptable?

Answer: _____

_____

_____

_____

Reference: _____

9. Does the *Code* either permit or prohibit a sprinkler head over a switchboard?

Answer: _____

_____

_____

_____

Reference: _____

10. A 2000-ampere, 3-phase, 480-volt ungrounded service is located 300 feet (91.4 m) from the water line within a building. Can the grounding electrode conductor from this service be connected to the building steel in the vicinity of the service, and then 300 feet (91.4 m) away bond the steel to the water line such that the building steel is being used as the grounding electrode conductor?

Answer: _____

_____

_____

_____

Reference: _____

11. Four 12 AWG THNN conductors in one conduit supply a continuous lighting load. What is the maximum permissible circuit breaker size and permissible load allowed on each conductor?

Answer: _____

_____

_____

_____

Reference: _____

12. A receptacle outlet is located in a soffit or roof overhang for connection of Christmas decorations or roof heating cable at a dwelling. When this receptacle is located out of reach from the grade, approximately 9 feet (2.75 m) above grade, is this receptacle required to be GFCI protected?

Answer: _____

_____

_____

_____

Reference: _____

13. Does the *NEC*® permit more than one Type NM cable under one device box cable clamp?

Answer: _____

_____

_____

_____

Reference: _____

14. A 400-ampere service uses two 200-ampere panelboards, each with a main breaker. Additional load is added so that one more 100-ampere service disconnect is required. Must the service-entrance conductors be increased in size? (The calculated load is not increased.)

Answer: _____

_____

_____

_____

Reference: _____

15. Can an attachment plug be used as a disconnecting means for a 3-horsepower motor?

Answer: _____

_____

_____

_____

Reference: _____

16. Where in the *Code* does it permit using 4 AWG copper THW conductors for service conductors supplying a 100-ampere service?

Answer: _____

_____

_____

_____

Reference: _____

17. In three-way and four-way switch circuitry in a metal raceway, is the neutral required to be in the same raceway with the travelers (switch legs)?

Answer: _____

_____

_____

_____

Reference: _____

18. Single-conductor UF cable comes underground from a submersible pump through a building wall to the pump controller. Are these conductors permitted to be run inside the building?

Answer: _____

_____

_____

_____

Reference: _____

19. In commercial garages, can regular receptacles and plates be used over 18 inches (450 mm) above the floor? What are the wiring restrictions in a garage that has a minimum of four air changes per hour?

Answer: _____

_____

_____

_____

Reference: _____

20. A single-family dwelling has two electric dryer receptacle outlets wired with two 10 AWG black conductors to each one and one 10 AWG white conductor neutral. Each outlet has a separate 30-ampere circuit breaker. Is one 10 AWG neutral conductor permitted to be used with this installation?

Answer: _____

_____

_____

_____

Reference: _____

21. Can trade size ⅜ (12) flexible metal conduit be used on machinery and boilers for limit switches, flow switches, can switches, and solenoids?

Answer: _____

_____

_____

_____

Reference: _____

22. Article 100—Definitions states that the definition of a building is *a structure that stands alone or that is cut off from adjoining structures by fire walls with all openings therein protected by approved fire doors.* Could this definition apply to a building with fire-rated floors and all elevator shafts and stairways protected by approved fire rating and fire doors? In other words, could each floor be considered a separate building?

Answer: _____

_____

_____

_____

Reference: _____

23. Can unit equipment used for emergency lighting be directly connected on the branch-circuit rather than use a plug-in receptacle?

Answer: _____

_____

_____

_____

Reference: _____

24. *NEC*® 517.20 specifies GFCI receptacles for wet locations that by definition in Article 100 include outside areas exposed to the weather. Are receptacles required to be GFCI protected when outdoors at health care facilities?

Answer: _____

_____

_____

_____

Reference: _____

25. Does a gas furnace in a dwelling require a disconnect switch on the furnace? May the furnace be directly connected onto a basement lighting circuit?

Answer: _____

_____

_____

_____

Reference: _____

PRACTICE EXAM 10

1. Are open tube fluorescent luminaires (fixtures) acceptable in a basement garage under an apartment building?

Answer: _____

_____

_____

_____

Reference: _____

2. Hot tubs are popular. Are there size and depth requirements to classify them as a pool or are they classified as a pool regardless of size for wiring of circulation pumps, heaters, and so forth?

Answer: _____

_____

_____

_____

Reference: _____

3. Rigid metal conduit runs in a commercial garage floor and extends up through an 18-inch (450 mm) hazardous area with no couplings or fittings in that 18-inch (450 mm) area. Is a seal-off fitting required in this conduit?

Answer: _____

_____

_____

_____

Reference: _____

4. Three single-phase, 480-volt transformers rated at 104 amperes each are connected in a wye bank for a 3-phase system. What is the maximum overcurrent protection permitted for the primary when the secondary is also protected? Would the primary protection be different if connected in a delta bank?

Answer: _____

_____

_____

_____

Reference: _____

5. How is the load for a section of multioutlet assembly (plug-mold) determined?

Answer: _____

_____

_____

_____

Reference: _____

6. A motor is connected to a branch-circuit breaker in a panel that is out of sight of the motor location. The breaker is used as the controller and also the disconnecting means for the motor. If this breaker is of the "lock off" type, would this meet *Code* requirements?

Answer: _____

_____

_____

_____

Reference: _____

7. A 12 AWG remote-control circuit is protected by 80-ampere fuses in the motor disconnect (magnetic contact). Is this control circuit properly protected?

Answer: _____

_____

_____

_____

Reference: _____

8. Bus duct is reduced in size from 1000 amperes to 400-ampere bus duct and runs 40 feet (12 m) to where it terminates in a distribution panel. Is overcurrent protection required?

Answer: _____

_____

_____

_____

Reference: _____

9. Can UF cable be used from the meter, down the pole, then underground 20 feet (6 m) to the disconnect for a pasture pump?

Answer: _____

_____

_____

_____

Reference: _____

10. On a 480/277-volt, 3-phase, 4-wire system where the neutral is not used, must it be grounded at the service panel?

Answer: _____

_____

_____

_____

Reference: _____

11. If available on the premises, the *Code* requires the bonding of all grounding electrodes together to form the grounding electrode system. Are these requirements the same for a separately derived system?

Answer: _____

_____

_____

_____

Reference: _____

12. Define tap conductors. Is it permissible to make a tap from a tap?

Answer: _____

_____

_____

_____

Reference: _____

13. Can two tandem circuit breakers mounted adjacent to each other in a panelboard with adjacent handles tied together be used in place of two double-pole breakers to supply two 220-volt electric baseboard heaters?

Answer: _____

_____

_____

_____

Reference: _____

14. What minimum size copper grounding electrode conductor can you use to ground a 400-ampere lighting service when two 3/0 AWG copper conductors in parallel are used for the phase conductors?

Answer: _____

_____

_____

_____

Reference: _____

15. Can *aluminum* rigid metal conduit be used in hazardous locations such as in sewer plants?

Answer: _____

_____

_____

_____

Reference: _____

16. A two-lamp exit luminaire (light) is supplied from both a battery-operated unit of equipment and a normal source. Must the normal supply to the exit luminaire (light) be from a recognized emergency source such as ahead of the service disconnect?

Answer: _____

_____

_____

_____

Reference: _____

17. A service is changed and moved to a different location. The old electric range run is Type SE cable that is too short to reach the new service, so a metal junction box is installed and a short piece of new Type SE cable is installed to reach the new service. This is in a residential basement on a 7-foot (2 m) ceiling. Does this metal junction box require grounding, and if so, does it require a separate grounding conductor to the service equipment (not the neutral of the 3-wire Type SE cable)?

    Answer: _____

    _____

    _____

    _____

    Reference: _____

18. Can service conductors be run directly to a fire pump controller?

    Answer: _____

    _____

    _____

    _____

    Reference: _____

19. A trucking firm has added a second service. They have 240-volt service to the old part of the building and 208-volt service to the new part. Common computer equipment is connected to both services with shielded TWINAX (signal cable) between devices. The two services are not tied to the same ground. This causes ground loop on signal shield. What is the best way per the *NEC®* to achieve equal potential bonding on network?

    Answer: _____

    _____

    _____

    _____

    Reference: _____

20. A luminaire (lighting fixture) is installed over a hydromassage bathtub. Is GFCI protection required for the luminaire (fixture)?

    Answer: _____

    _____

    _____

    _____

    Reference: _____

21. Frequently the *NEC*® requires GFCI protection of receptacles. Is there any requirement that limits the number of receptacles that can be protected by a single GFCI circuit breaker or a feed-through receptacle?

Answer: _____

_____

_____

_____

Reference: _____

22. Can an electrical discharge luminaire (fixture) such as a fluorescent strip be used as a branch-circuit junction box or termination point for circuit conductors?

Answer: _____

_____

_____

_____

Reference: _____

23. Does the *Code* allow electrical nonmetallic tubing (ENT) to be installed above a suspended ceiling provided that a thermal barrier of material has at least a 15-minute rating of fire-rated assemblies?

Answer: _____

_____

_____

_____

Reference: _____

24. Does the *Code* permit 8 AWG conductors to be installed in a 4 × 2⅛ inch-square nonmetallic box that is marked with the number of 14 AWG, 12 AWG, and 10 AWG conductors that can be installed in it?

Answer: _____

_____

_____

_____

Reference: _____

25. Must lay-in fluorescent luminaires (lighting fixtures) be fastened or can they lay in grid T-bars of a suspended ceiling system?

Answer: _____

_____

_____

_____

Reference: _____

PRACTICE EXAM 11

1. When Type UF cable is used for interior wiring as permitted, are the conductors required to be rated at 90°C?

   Answer: _____

   _____

   _____

   _____

   Reference: _____

2. Are swimming pool underwater luminaires (lighting fixtures) operating at less than 15 volts between conductors required to be protected by a GFCI device or circuit breaker?

   Answer: _____

   _____

   _____

   _____

   Reference: _____

3. Can handle locks on circuit breakers be locked so that the power to loads such as emergency lighting, sump pumps, alarm warning circuits, and other types of equipment cannot be cut off by mistake? If so, what sections of the *NEC*® would apply?

   Answer: _____

   _____

   _____

   _____

   Reference: _____

4. Can nonmetallic-sheathed cable be run parallel to framing members through cold air returns?

   Answer: _____

   _____

   _____

   _____

   Reference: _____

5. Is a GFCI properly protected circuit supplying an outdoor portable sign an acceptable method for installation in accordance with the *NEC*®?

   Answer: _____

   _____

   _____

   _____

   Reference: _____

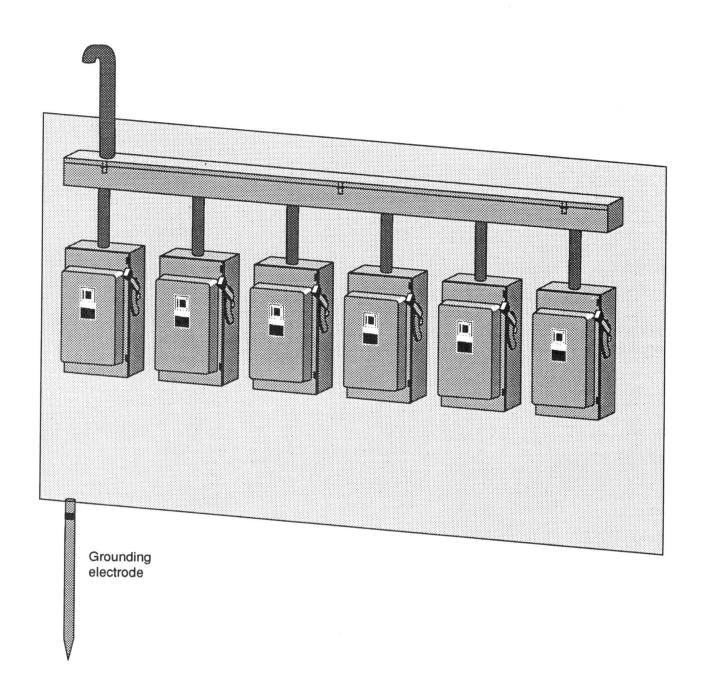

Grounding
electrode

6. *Refer to the diagram above*. Where should the grounding electrode conductor terminate in a multiple disconnect service?

Answer: _____

_____

_____

_____

Reference: _____

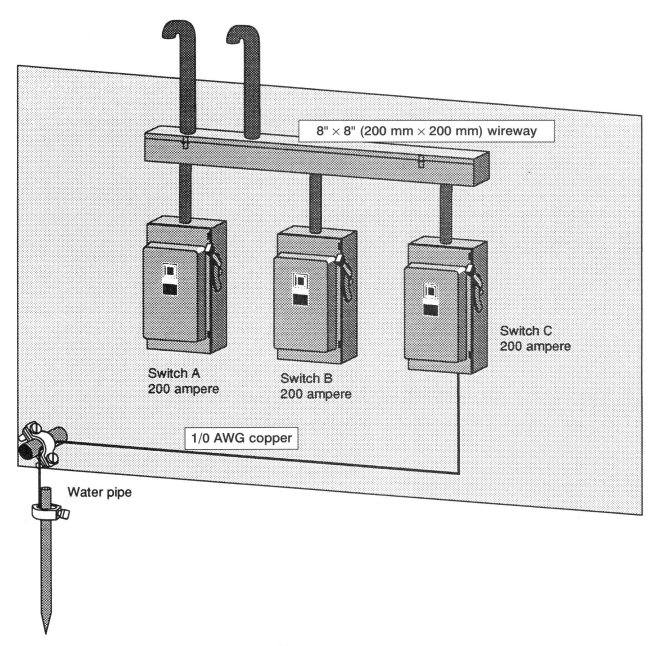

Refer to the diagram above for Questions 7 and 8.

7. A 600-ampere service is built with parallel sets of 350-kcmil CU THWN service conductors. Three 200-ampere fused service disconnects are supplied with 4/0 AWG THWN service taps within an 8" × 8" (200 mm × 200 mm) wireway. All service bonding is continuous from the CT cabinet through all service enclosures. Does the *Code* permit only one grounding electrode conductor connection at switch "C" if sized for 700-kcmil service conductors?

Answer: _____

_____

_____

_____

Reference: _____

8. With reference to Question 7, how should the trough above the three service disconnects be sized? Is the trough to be sized as a wireway or a junction box?

Answer: _____

_____

_____

_____

Reference: _____

9. An EMT and steel box system was installed throughout an office/warehouse building. An additional equipment grounding conductor was pulled with all receptacle branch-circuits and terminated at all receptacle grounding screws. Must the equipment grounding conductor also be attached to each metal box?

Answer: _____

_____

_____

_____

Reference: _____

10. Is an insulated equipment grounding conductor required in nonmetallic sheathed cable that is supplying a swimming pool pump motor installed in the garage of a single-family dwelling?

Answer: _____

_____

_____

_____

Reference: _____

11. A 225-ampere rated power panel has main lugs only and powers a calculated load of 180 amperes. It is supplied by a 500-kcmil CU feeder with 400-ampere overcurrent protection. Does this installation comply with the *Code*?

Answer: _____

_____

_____

_____

Reference: _____

12. If a small dental office uses limited amounts of inhalation anesthetics, does *NEC*® 517.31 require standby emergency power to supply emergency lighting and the anesthetic equipment?

Answer: _____

_____

_____

_____

Reference: _____

13. Does the *Code* allow any derating of loads in a commercial laundry? The normal load of a washing machine is 4 amperes; the spin cycle draws 9 amperes. Because it is unlikely that all machines will be spinning at the same time, what load figure should be used for load calculations?

Answer: _____

_____

_____

_____

Reference: _____

14. Is the examining room of a dental office considered to be a general-care patient care area? Is Type NM cable a permitted wiring method in this area?

Answer: _____

_____

_____

_____

Reference: _____

15. Do the metal enclosures for individual channel-type neon letters (either front-lit or back-lit) require grounding?

Answer: _____

_____

_____

_____

Reference: _____

16. Is it permissible to ground the secondary of a separately derived system back to the neutral terminal of a service switchboard instead of going directly to the water line when there is no effectively grounded structural steel nearby?

Answer: _____

_____

_____

_____

Reference: _____

17. When conductors are paralleled, does the ampacity double (assuming the same type, size, length)?

Answer: _____

_____

_____

_____

Reference: _____

18. Suppose you use three 300-kcmil conductors in a raceway in free air with 75°C temperature rating of 285 amperes for each conductor. Instead of this, can you use six 1/0 AWG conductors with 75°C rating and two conductors in parallel per phase, all in the same conduit at 300 amperes ampacity, or must you derate because you have more than three conductors? (Assume that you are using Type THWN conductors in a trade size 2½ (63) conduit with device terminations rated for 75°C.)

Answer: _____

_____

_____

_____

Reference: _____

19. Are medium-base HID lampholders permitted to be installed on a 277-volt circuit?

Answer: _____

_____

_____

_____

Reference: _____

20. Is GFCI protection required for receptacles located on decks where the receptacles are located less than 6 feet, 6 inches (2 m) above grade and access to the deck is by steps or stepping onto a low deck? (The receptacle cannot be reached while standing on the ground.)

Answer: _____

_____

_____

_____

Reference: _____

21. Are bathroom GFCI receptacles permitted to supply luminaire (lighting) outlets because tripping of the GFCI causes the lights to go out?

Answer: _____

_____

_____

_____

Reference: _____

22. Is it acceptable to connect a garbage disposal and dishwasher on the same branch-circuit?

    Answer: _____

    _____

    _____

    _____

    Reference: _____

23. Are receptacles required in a four-car garage at an apartment building if the garage is detached and provided with electric power and lights?

    Answer: _____

    _____

    _____

    _____

    Reference: _____

24. Does an outdoor entrance luminaire (lighting fixture) with integral photocell satisfy the *Code* without a wall switch control? What about a motion detector used to activate the luminaire (lighting fixture)?

    Answer: _____

    _____

    _____

    _____

    Reference: _____

25. Does the *Code* require feeder conductors to have an ampacity of 125 percent of the continuous load plus the noncontinuous load, or is it only necessary for the feeder overcurrent device to have this rating? If only the feeder overcurrent device requires this rating, can the feeder conductors be sized only for total continuous plus noncontinuous load?

    Answer: _____

    _____

    _____

    _____

    Reference: _____

PRACTICE EXAM 12

1. In calculating feeder loads for electric space heating, is it required to figure 125 percent of space heating loads or is the 125 percent required only for branch-circuit loads?

Answer: _____

_____

_____

_____

Reference: _____

2. For household range load calculations, is it acceptable to combine ovens and countertops and treat them as one appliance for feeder calculations?

Answer: _____

_____

_____

_____

Reference: _____

3. Because the disconnecting means for separate buildings on one property are required to be suitable for use as service equipment, are the number of disconnects at each building's entrance limited to six?

Answer: _____

_____

_____

_____

Reference: _____

4. The *Code* requires multiple services to be grounded to the same electrode. Must the grounding connections be made to the same point on the electrode or can they be grounded to the same electrode (water line or structural steel) at different locations?

Answer: _____

_____

_____

_____

Reference: _____

5. Does Article 514 apply to underground wiring (such as to a sign) that is at least 30 feet (9 m) from any hazardous location in a service station?

Answer: _____

_____

_____

_____

Reference: _____

6. What is the required minimum clearance between a thermally protected Type IC recessed incandescent luminaire (fixture) and wood framing?

Answer: _____

_____

_____

_____

Reference: _____

7. Does the *Code* permit the metal coaxial sheath of CATV cable to be grounded to a separate driven rod or is it required to be bonded to the building's power system ground or service equipment enclosure?

Answer: _____

_____

_____

_____

Reference: _____

8. An automotive repair garage is classified as Class I, Division 2 up 18 inches (450 mm) above the floor. An attached sales and office area is not classified. Is Type NM-B cable permitted to be used in the sales and office area?

Answer: _____

_____

_____

_____

Reference: _____

9. Do Type AC and MC cables installed on 277-volt circuits require bonding if connected into concentric knockouts?

Answer: _____

_____

_____

_____

Reference: _____

10. A swimming pool is installed at a single-family dwelling. Can 2-wire Type NM-B cable with a bare grounding conductor be used to connect the pump motor located in the basement?

Answer: _____

_____

_____

_____

Reference: _____

11. Ceiling lighting consists of lay-in fluorescent luminaires (fixtures). Is it acceptable to use trade size ⅜ (12) flexible metal conduit from luminaire to luminaire (fixture to fixture) in lengths less than 6 feet (1.8 m) provided an equipment grounding conductor is installed in the conduit?

Answer: _____

_____

_____

_____

Reference: _____

12. Can a 12-volt dry-niche luminaire (lighting fixture) mounted on the outside of a permanent aboveground swimming pool be supplied by flexible cord connected to a receptacle located 11 feet (3.3 m) from the pool?

Answer: _____

_____

_____

_____

Reference: _____

13. Can a timer switch for a spa be located within 5 feet (1.5 m) of the spa? What if it is located on the unit and not on the wall?

Answer: _____

_____

_____

_____

Reference: _____

14. Is there any requirement relative to the location of the wall switch for an outdoor entrance luminaire (light)? As an example, must the switch for an entrance luminaire (light) at a sunroom door be located at the entrance door from outside or can it be located 10 feet (3 m) away at another door leading to the sunroom?

Answer: _____

_____

_____

_____

Reference: _____

15. When calculating the demand for a 3-kW and a 6-kW range, would you use Table 220.55, Column B for the 3-kW and Column C for the 6-kW or would you combine them and use either B or C for the two ranges?

Answer: _____

_____

_____

_____

Reference: _____

16. Can more than one receptacle in a laundry room be supplied by the laundry branch-circuit?

    Answer: _____

    _____

    _____

    _____

    Reference: _____

17. For a large bathroom with provisions for a washing machine, is GFCI protection required for an inaccessible receptacle located behind and dedicated to the washing machine?

    Answer: _____

    _____

    _____

    _____

    Reference: _____

18. Is there an acceptable method whereby control wiring can be run in the same conduit with power conductors to a central air-conditioning unit?

    Answer: _____

    _____

    _____

    _____

    Reference: _____

19. Does the *Code* require grounding the metal sheath of CATV cable entering a dwelling?

    Answer: _____

    _____

    _____

    _____

    Reference: _____

20. The *Code* requires bonding a metallic conduit enclosing a grounding electrode conductor. How is the bonding jumper sized when the conduit enters a concentric knockout?

    Answer: _____

    _____

    _____

    _____

    Reference: _____

21. Does the *Code* require a GFCI-protected receptacle when installed within 6 feet (1.8 m) of a dwelling wet bar or laundry sink?

    Answer: _____

    _____

    _____

    _____

    Reference: _____

22. A 500-kcmil Type THW CU feeder protected with 400-ampere overcurrent device is routed through a junction box. What is the minimum size tap conductor required to supply a 30-ampere load using the 25-foot (7.5 m) tap rule? What is the minimum size using the 10-foot (3 m) tap rule?

    Answer: _____

    _____

    _____

    _____

    Reference: _____

23. A large recreation room in a dwelling has three sets of sliding doors opening onto a covered porch. Is a switch required at each door to control porch lumination (lighting)?

    Answer: _____

    _____

    _____

    _____

    Reference: _____

24. Existing feeder conduits are being repulled using conductors with 90°C Type THHN insulation. What ampacity values are allowed for these conductors that are connected to a circuit breaker or panel main lug terminals?

    Answer: _____

    _____

    _____

    _____

    Reference: _____

25. Branch-circuits serving patient care areas must be installed in metal raceways. Which areas in nursing homes and residential custodial care facilities are designated as patient care areas?

    Answer: _____

    _____

    _____

    _____

    Reference: _____

FINAL EXAMINATION

Examination Instructions:

For a positive evaluation of your knowledge and preparation awareness, you must:

1. locate yourself in a quiet atmosphere (room by yourself).

2. have with you at least two sharp No. 2 pencils, the 2005 *NEC®*, and a handheld calculator.

3. time yourself (three hours) with no interruptions. Do not spend more than 3½ minutes per question.

4. after the test is complete, grade yourself honestly and concentrate your studies on the sections of the *NEC®* in which you missed the questions.

Caution: Do not just look up the correct answers, because the questions in this examination are only an exercise and not actual test questions. Therefore, it is important that you be able to quickly find answers from throughout the *NEC®*.

1. According to the *National Electrical Code®*, open conductors for communication equipment on a building shall be separated at least _____ feet (_____ m) from *lightning* conductors.
   A. 2 (600 mm)
   B. 4 (1.2)
   C. 6 (1.8)
   D. 8 (2.4)

   Answer: _____   Reference: _____

2. A megohmmeter is an instrument used for
   A. polarizing a circuit.
   B. measuring high resistances.
   C. shunting a generation system.
   D. determining amperes.

   Answer: _____   Reference: _____

3. The total opposition to alternating current in a circuit that includes resistance, inductance, and capacitance is called
   A. reactance.
   B. resistance.
   C. reluctance.
   D. impedance.

   Answer: _____   Reference: _____

4. A 240-volt, single-phase circuit has a resistive load of 8500 watts. The net calculated current to supply this load is _____ amperes.
   A. 35
   B. 39
   C. 44
   D. 71

   Answer: _____   Reference: _____

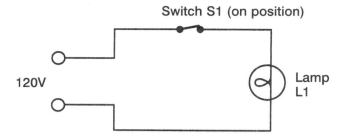

Refer to the diagram above for Questions 5 and 6.

5. In the diagram above, switch S1 is in the "on" position, but luminaire (light) L1 does not come on. Voltage across L1 is measured to be 120 volts. Voltage across S1 is measured to be 0 volt. The luminaire (light) does not come on because
   A. the luminaire (light) is open (burned out).
   B. the luminaire (light) and switch are shorted.
   C. the luminaire (light) is good but the switch does not make contact.
   D. there is a break in the wire of the circuit.

   Answer: _____   Reference: _____

6. In the diagram above, with a 9-ampere current in the circuit, the power factor is _____ percent.
   A. 71
   B. 83
   C. 93
   D. 108

   Answer: _____   Reference: _____

480V 3-phase

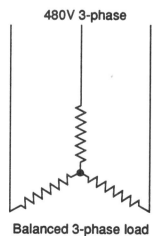

Balanced 3-phase load

7. In the diagram above, three balanced resistive 100-ampere loads are connected to a 480-volt, 3-phase, 3-wire circuit. The total power of this circuit is _____ kilowatts.
   A. 48
   B. 72
   C. 83
   D. 144

   Answer: _____   Reference: _____

8. To get the maximum total resistance using three resistors,
   A. all three resistors should be connected in series.
   B. all three resistors should be connected in parallel.
   C. two resistors should be connected in parallel then one connected in series.
   D. two resistors should be connected in series then one connected in parallel.

   Answer: _____ Reference: _____

9. Four resistance heaters are connected in parallel. Their resistances are heater 1, 20 ohms; heater 2, 30 ohms; heater 3, 60 ohms; and heater 4, 10 ohms. The total resistance of the parallel circuit is _____ ohms.
   A. 5
   B. 12
   C. 50
   D. 120

   Answer: _____ Reference: _____

10. A building on a blueprint is 16 inches × 10 inches. If the drawing scale is ¼ inch = 1 foot, what is the area of the building in square feet?
    A. 160 square feet
    B. 640 square feet
    C. 2560 square feet
    D. 5120 square feet

    Answer: _____ Reference: _____

The following diagram is for Question 11.

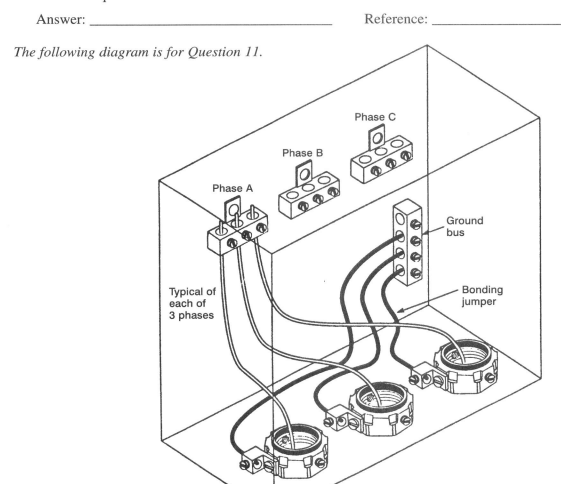

11. In the diagram on page 293, each of the supply-side EMT conduits contains three 500-kcmil copper THW service conductors in parallel. As shown, a separate bonding jumper is installed from each conduit to the grounded bus terminal. Each bonding jumper must be at least _____ AWG copper.
    A. 1/0
    B. 2/0
    C. 3/0
    D. 4/0

    Answer: _____   Reference: _____

12. A single-phase, 3-wire service has two ungrounded conductors of 2/0 AWG copper THWN. The neutral conductor is 1 AWG copper THWN. The conduit size needed for the service-entrance conductors is at least trade size _____ (_____).
    A. 1½ (41)
    B. 2 (53)
    C. 2½ (63)
    D. 3 (78)

    Answer: _____   Reference: _____

13. A metallic cold water piping system is available within the structure being supplied by an electrical service with 1/0 AWG THW copper service-entrance conductors. The copper grounding-electrode conductors run to the metal water pipe shall have a minimum size of _____ AWG.
    A. 8
    B. 6
    C. 4
    D. 2

    Answer: _____   Reference: _____

14. Ground-fault protection is required to be installed on a 2000-ampere, solidly grounded, wye service. What is the maximum setting of the ground fault protection?
    A. 800 amperes
    B. 1000 amperes
    C. 1200 amperes
    D. 1600 amperes

    Answer: _____   Reference: _____

15. Overhead service conductors shall be installed so that the minimum clearance from a window opening is _____ feet (_____ mm/m).
    A. 2 (600 mm)
    B. 3 (900 mm)
    C. 6 (1.8 m)
    D. 7 (2.4 m)

    Answer: _____   Reference: _____

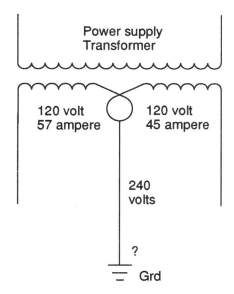

16. The 240-volt, single-phase feeder shown in the diagram above supplies a branch-circuit panel that has an improperly balanced load of 57 amperes on one ungrounded conductor and 45 amperes on the other ungrounded conductor. What is the load on the grounded (neutral) conductor?
    A. no load
    B. 12 amperes
    C. 57 amperes
    D. 102 amperes

    Answer: _____ Reference: _____

17. The maximum number of power or lighting conductors that can be installed in a raceway before the derating factors must be applied is
    A. 1.
    B. 3.
    C. 5.
    D. 7.

    Answer: _____ Reference: _____

18. A 230-volt, single-phase, 100-ampere circuit is installed in a nonmetallic raceway. Which of the following conditions must apply to the equipment grounding conductor installed with the circuit?
    I.   It must be counted when determining conductor fill of the raceway.
    II. It must be the same size as that of the circuit conductor.
    A. I only
    B. II only
    C. both I and II
    D. neither I nor II

    Answer: _____ Reference: _____

19. Three 4 AWG THWN and four 1/0 AWG THW conductors are to be installed in a single run of IMC conduit. The minimum trade size of conduit permitted is
    A. 1 (27).
    B. 2 (53).
    C. 3 (78).
    D. 4 (103).

    Answer: _____   Reference: _____

20. Eight 4 AWG THWN conductors are to be installed in a single run of rigid nonmetallic conduit. The minimum trade size of conduit permitted is
    A. 1 (27).
    B. 1½ (41).
    C. 2 (53).
    D. 2½ (63).

    Answer: _____   Reference: _____

21. A service disconnecting means can be installed at which of the following locations?
    I. Outside a building at a readily accessible point nearest the point of entrance of the service-entrance conductors.
    II. Inside a building at a readily accessible point nearest the point of entrance of the service-entrance conductors.
    A. I only
    B. II only
    C. either I or II
    D. neither I nor II

    Answer: _____   Reference: _____

Conductors crossing and connecting

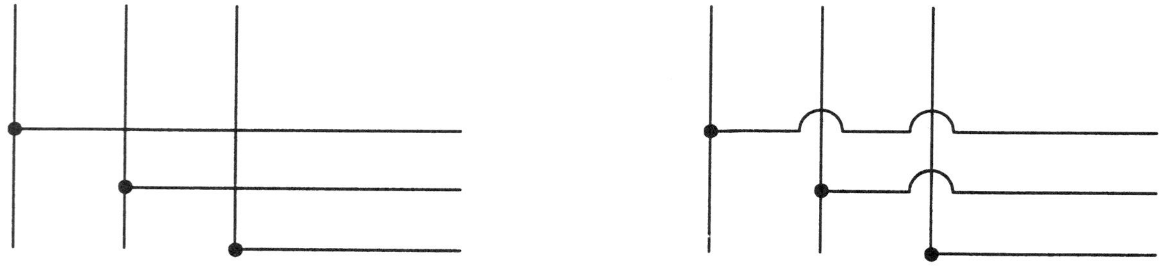

Figure 1   Figure 2

22. Figure 1 and Figure 2 shown above are standard electrical symbols used on blueprints representing crossing conductors. Which symbol represents two conductors crossing but not connecting?
    A. Figure 1 only
    B. Figure 2 only
    C. Figure 1 and Figure 2
    D. Neither Figure 1 nor Figure 2

    Answer: _____   Reference: _____

23. The following figure is a standard symbol used on blueprints. The symbol represents which of the following?
    A. an air circuit breaker
    B. a lightning arrester
    C. a fuse
    D. a thermal element

    Answer: _____   Reference: _____

24. A motor is protected against short-circuit and ground-fault by an adjustable instantaneous trip circuit breaker that is part of a combination controller having motor overload and short-circuit and ground-fault protection in each conductor. The setting of the instantaneous trip breaker shall be permitted to exceed the motor full-load current by not more than _____ percent.
    A. 250
    B. 1300
    C. 700
    D. 1000

    Answer: _____   Reference: _____

25. Thermal overload relays are used for the protection of polyphase induction motors. Their primary purpose is to protect the motor in case of
    A. reversal of phases in the supply.
    B. low line voltage.
    C. short-circuit between phases.
    D. sustained overload.

    Answer: _____   Reference: _____

26. A motor controller and motor branch-circuit disconnecting means a 2300-volt motor shall have a continuous ampere rating of not less than the
    A. trip setting of the short-circuit protective device rating.
    B. trip setting of the overload protection device.
    C. trip setting of the fault-current protection device.
    D. locked-rotor rating of the motor.

    Answer: _____   Reference: _____

27. The disconnecting means for both a 2300-volt motor and the controller shall be
    A. permitted within the same enclosure as the controller.
    B. located separately from the controller enclosure.
    C. located at the service equipment.
    D. permitted within the watt-hour meter enclosure.

    Answer: _____   Reference: _____

28. Speed-limiting devices shall be provided with which of the following?
    A. polyphase squirrel-cage motors
    B. synchronous motors
    C. compound motors
    D. series motors

    Answer: _____   Reference: _____

29. Which of the following appliances can be grounded to the grounded (neutral) conductor in an existing dwelling?
    A. electric water heater
    B. kitchen disposal
    C. dishwasher
    D. electric dryer

    Answer: _____   Reference: _____

30. A metal luminaire (lighting fixture) shall be grounded if located
    A. 10 feet (3 m) vertically or 6 feet (1.8 m) horizontally from a kitchen sink.
    B. 8 feet (2.4 m) vertically or 5 feet (1.5 m) horizontally from a kitchen sink.
    C. 6 feet (1.8 m) vertically or 3 feet (900 mm) horizontally from a kitchen sink.
    D. 8 feet (2.4 m) vertically or 3 feet (900 mm) horizontally from a kitchen sink.

    Answer: _____   Reference: _____

31. When used, a driven ground rod shall be installed so that the soil will be in contact with a length of the rod not less than
    A. 4 feet (1.2 m).
    B. 6 feet (1.8 m).
    C. 8 feet (2.4 m).
    D. 10 feet (3 m).

    Answer: _____   Reference: _____

32. Which of the following most accurately describes the condition of a motor known as "locked-rotor?"
    A. When the electrician places a lock on the motor controller to keep the motor from being energized
    B. Mechanical brakes used on the motor shaft to stop the motor during shutdown
    C. An electronic control device used to lock the speed of the motor at that specified by the manufacturer
    D. When the circuits of a motor are energized but the rotor is not turning

    Answer: _____   Reference: _____

33. The following diagram represents overhead conductors between two buildings on an industrial site. The voltage is 240/480 volts ac. The conductors pass over a driveway leading to a loading dock at one of the buildings. What is the minimum vertical clearance permitted between the overhead conductors and the driveway?

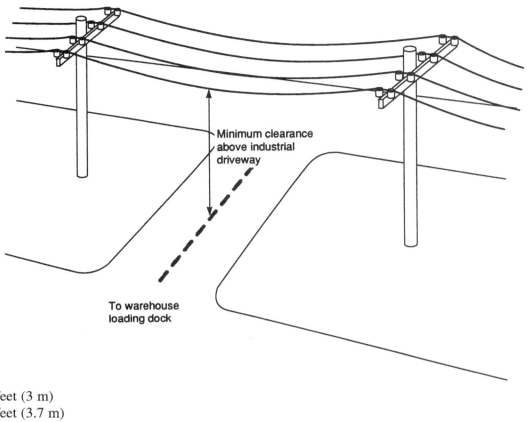

Minimum clearance above industrial driveway

To warehouse loading dock

A. 10 feet (3 m)
B. 12 feet (3.7 m)
C. 15 feet (4.5 m)
D. 18 feet (5.5 m)

Answer: _____ Reference: _____

34. When a ground-fault protection for equipment is installed within service equipment, it shall be performance tested
A. at the factory before shipment.
B. before being installed on-site.
C. when first installed on-site.
D. after the electrical system has been used for one week.

Answer: _____ Reference: _____

35. Three continuous-duty motors with full-load current ratings of 5.6 amperes, 4.5 amperes, and 4.5 amperes, respectively, are to be installed on a single branch-circuit. The circuit conductors shall have a minimum ampacity of _____ amperes.
A. 16
B. 20
C. 24
D. 30

Answer: _____ Reference: _____

36. The following diagram represents an electrical service with a fused switch as the main disconnect. What is the maximum height the center of the grip of the operating handle of the switch is permitted to be located above the ground?

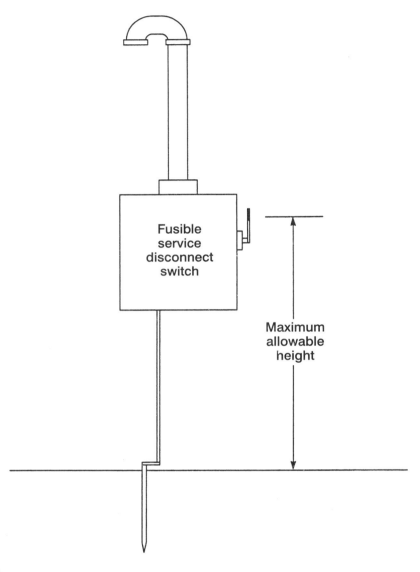

A. 5 feet (1.5 m)
B. 6 feet 7 inches (2 m)
C. 7 feet (2.1 m)
D. ½ foot (150 mm)

Answer: _____ Reference: _____

37. How is the size of an electrical conductor supplying a circuit determined?
    A. voltage
    B. amperage
    C. the length
    D. all of the above

Answer: _____ Reference: _____

38. Ohm's law is
    A. measurement of the I²R losses.
    B. the relationship between voltage, current, and power.
    C. an equation for determining power.
    D. a relationship between voltage, current, and resistance.

    Answer: _____   Reference: _____

39. Electrical pressure is measured in
    A. volts.
    B. amperes.
    C. coulombs.
    D. watts.

    Answer: _____   Reference: _____

40. The resistance of a 1500-watt, 120-volt resistance heater element is
    A. 14.4 ohms.
    B. 9.6 ohms.
    C. 11.2 ohms.
    D. 12.5 ohms.

    Answer: _____   Reference: _____

41. Total ampacity of stranded or solid conductors is the same if they have the same
    A. diameter.
    B. circumference.
    C. cross-sectional area.
    D. insulation.

    Answer: _____   Reference: _____

42. A branch-circuit that supplies a number of outlets for lighting and appliances is called a _____ branch-circuit.
    A. general-purpose
    B. utility
    C. multipurpose
    D. none of the above

    Answer: _____   Reference: _____

43. A 10-ohm resistance carrying 10 amperes of current uses _____ watts of power.
    A. 100
    B. 200
    C. 500
    D. 1000

    Answer: _____   Reference: _____

44. A _____ stores energy in much the same manner as a spring stores mechanical energy.
    A. coil
    B. capacitor
    C. resistor
    D. none of the above

    Answer: _____   Reference: _____

45. _____ means that it is constructed or protected so that exposure to the weather will not interfere with its successful operation.
    A. Weatherproof
    B. Weathertight
    C. Weather resistant
    D. all of the above

    Answer: _____ Reference: _____

46. The service conductor clearance from windows can be less than 3 feet if it is
    A. not less than 24 inches (600 mm) from the window.
    B. run along the left side of the window.
    C. run below the bottom level of the window.
    D. none of the above.

    Answer: _____ Reference: _____

47. The NEC® states that electrical equipment shall be installed
    A. not exceeding the provisions of the Code.
    B. not less than the Code permits.
    C. according to the Code and local code amendments.
    D. none of the above

    Answer: _____ Reference: _____

48. The equivalent resistance of three resistors of 8 ohms, 8 ohms, and 4 ohms that are connected in parallel is
    A. 2 ohms.
    B. 4 ohms.
    C. 6 ohms.
    D. 8 ohms.

    Answer: _____ Reference: _____

49. Because fuses are rated by both amperage and voltage, a fuse will operate correctly on
    A. ac only.
    B. ac or dc.
    C. dc only.
    D. any voltage.

    Answer: _____ Reference: _____

50. The following 480-volt, 3-phase, 3-wire, intermittent use equipment is in a commercial kitchen: two 5000-watt water heaters; four 3000-watt fryers; and two 6000-watt ovens. Each ungrounded conductor in the feeder circuit for this kitchen equipment must be sized to carry a minimum calculated load of _____ amperes.
    A. 15
    B. 27
    C. 41
    D. 46

    Answer: _____ Reference: _____

Appendix 1: Symbols

Caution: These are standard symbols. Some variation is common from region to region.

(Courtesy of American Iron and Steel Institute)

Electrical Wiring Symbols

Selected from American National Standard Graphic for
Electrical Wiring and Layout Diagrams Used in Architecture and Building Construction

ANSI Y32.9-1972

1. Lighting Outlets

Ceiling　　　　　　　　　　　　　　*Wall*

1.1 Surface or pendant incandescent, mercury-vapor, or similar lamp fixture

1.2 Recessed incandescent, mercury-vapor, or similar lamp fixture

1.3 Surface of pendant individual fluorescent fixture

1.4 Recessed individual fluorescent fixture

1.5 Surface or pendant continuous-row fluorescent fixture

1.6 Recessed continuous-row fluorescent fixture

1.8 Surface or pendant exit light

1.9 Recessed exit light

1.10 Blanked outlet

1.11 Junction box

1.12 Outlet controlled by low-voltage switching when relay is installed in outlet box

2. Receptacle Outlets

Grounded　　　　　　　　　　*Ungrounded*

2.1 Single receptacle outlet

2.2 Duplex receptacle outlet

2.3 Triplex receptacle outlet

2.4 Quadrex receptacle outlet

2.5 Duplex receptacle outlet – split wired

2.6 Triplex receptacle outlet – split wired

2.7 Single special-purpose receptacle outlet -- split wired

NOTE 2.7A: Use numeral or letter as a subscript alongside the symbol, keyed to explanation in the drawing list of symbols, to indicate type of receptacle or use

2.8 Duplex special-purpose receptacle outlet
　See note 2.7A

2.9 Range outlet (typical)
See note 2.7A

2.10 Special-purpose connection or provision for connection
Use subscript letters to indicate function
(SW – dishwasher; CD – clothes dryer, etc).

DW   UNG DW

2.12 Clock hanger receptacle

C   C UNG

2.13 Fan hanger receptacle

F   F UNG

2.14 Floor single receptacle outlet

UNG

2.15 Floor duplex receptacle outlet

UNG

2.16 Floor special-purpose outlet
See note 2.7A

UNG

2.17 Floor telephone outlet – public

2.18 Floor telephone outlet – private

2.19 Underfloor duct and junction box for triple, double, or single duct system (as indicated by the number of parallel lines)

2.20 Cellular floor header duct

3. Switch Outlets

3.1 Single-pole switch
S

3.2 Double-pole switch
S_2

3.3 Three-way switch
S_3

3.4 Four-way switch
S_4

3.5 Key-operated switch
S_K

3.6 Switch and pilot lamp
S_P

3.7 Switch for low-voltage switching system
S_L

3.8 Master switch for low-voltage switching system
S_{LM}

3.9 Switch and single receptacle
$-\bigcirc S$

3.10 Switch and double receptacle
$=\bigcirc S$

3.11 Door switch
S_D

3.12 Time switch
S_T

3.13 Circuit breaker switch
S_{CB}

3.14 Momentary contact switch or pushbutton for other than signaling system

SMC

3.15 Ceiling pull switch

(S)

5.13 Radio outlet

[R]

5.14 Television outlet

[TV]

6. Panelboards, switchboards, and related equipment

6.1 Flush-mounted panelboard and cabinet
NOTE 6.1A: Identify by notation or schedule

6.2 Surface-mounted panelboard and cabinet
See note 6.1A

6.3 Switchboard, power control center, unit substations (should be drawn to scale)
See note 6.1A

6.4 Flush-mounted terminal cabinet
See note 6.1A

NOTE 6.4A: In small-scale drawings theTC may be indicated alongside the symbol

TC

6.5 Surface-mounted terminal cabinet
See note 6.1A and 6.4A

TC

6.6 Pull box
Identify in relation to wiring system section and size

6.7 Motor or other power controller
See note 6.1A

MC

6.8 Externally operated disconnection switch
See note 6.1A

6.9 Combination controller and disconnection means
See note 6.1A

7. Bus Ducts and Wireways

7.1 Trolley duct
See note 6.1A

| T | | T | | T |

7.2 Busway (service, feeder, or plug-in)
See note 6.1A

| B | | B | | B |

7.3 Cable through, ladder, or channel
See note 6.1A

| BP | | BP | | BP |

7.4 Wireway
See note 6.1A

| W | | W | | W |

9. Circuiting

Wiring method identification by notation on drawing or in specifications.

9.1 Wiring concealed in ceiling or wall

NOTE 9.1A: Use heavyweight line to identify service and feed runs.

9.2 Wiring concealed in floor
See note 9.1A

Appendix 2: Basic Electrical Formulas

Basic Electrical Formulas
(Courtesy of American Iron and Steel Institute)

DC Circuit Characteristics

Ohm's Law:

$$E = IR \quad I = \frac{E}{R} \quad R = \frac{E}{I}$$

E = voltage impressed on circuit (volts)
I = current flowing in circuit (amperes)
R = circuit resistance (ohms)

Resistances in Series:

$R_t = R_1 + R_2 + R_3 + \ldots$

R_T = total resistance (ohms)
R_1, R_2 etc. = individual resistances (ohms)

Resistances in Parallel:

$$R_t = \frac{1}{\frac{1}{R_1} + \frac{1}{R_2} + \frac{1}{R_3} + \ldots}$$

Formulas for the conversion of electrical and mechanical power:

$$HP = \frac{watts}{746} (watts \times .00134)$$

$$HP = \frac{kilowatts}{.746} (kilowatts \times 1.34)$$

Kilowatts = HP × .746
Watts = HP × 746
HP = Horsepower

In direct-current circuits, electrical power is equal to the product of the voltage and current:

$$P = EI = I_2 R = \frac{E_2}{R}$$

P = power (watts)
E = voltage (volts)
I = current (amperes)
R = resistance (ohms)

Solving the basic formula for I, E, and R gives

$$I = \frac{P}{E} = \sqrt{\frac{P}{R}}; \quad E = \frac{P}{I} = \sqrt{RP}; \quad R = \frac{E2}{P} = \frac{P}{T^2}$$

Energy

Energy is the capacity for doing work. Electrical energy is expressed in kilowatt-hours (kWhr), one kilowatt-hour representing the energy expended by a power source of 1 kW over a period of 1 hour.

Efficiency

Efficiency of a machine, motor or other device is the ratio of the energy output (useful energy delivered by the machine) to the energy input (energy delivered to the machine), usually expressed as a percentage:

$$Efficiency = \frac{output}{input} \times 100\%$$

$$or \; Output = Input \times \frac{efficiency}{100\%}$$

Torque

Torque may be described as a force tending to cause a body to rotate. It is expressed in pound-feet or pounds of force acting at a certain radius:

Torque (pound-feet) = force tending to produce rotation (pounds) × distance from center of rotation to point at which force is applied (feet).

Relations between torque and horsepower:

$$Torque = \frac{33,000 \times HP}{6.28 \times rpm}$$

$$HP = \frac{6.28 \times rpm \; time \; torque}{33,000}$$

rpm = speed of rotating part (revolutions per minute)

AC Circuit Characteristics

The instantaneous values of an alternating current or voltage vary from zero to maximum value each half cycle. In the practical formula that follows, the "effective value" of current and voltage is used, defined as follows:

Effective value = 0.707 × maximum instantaneous value

Inductances in Series and Parallel:

The resulting circuit inductance of several inductances in series or parallel is determined exactly as the sum of resistances in series or parallel as described under dc circuit characteristics.

Impedance:

Impedance is the total opposition to the flow of alternating current. It is a function of resistance, capacitive reactance and inductive reactance. The following formulae relate these circuit properties:

$$X_L = 2\pi Hz L \quad X_c = \frac{1}{2\pi Hz C} \quad Z = \sqrt{R^2 + (X_L - X_C^2)}$$

X_L = inductive reactance (ohms)
X_c = capacitive reactance (ohms)
Z = impedance (ohms)
Hz - (Hertz) cycles per second
C = capacitance (farads)

L - inductance (henrys)
R = resistance (ohms)
π = 3.14

In circuits where one or more of the properties L, C, or R is absent, the impedance formula is simplified as follows:

Resistance only: Inductance only: Capacitance only:
$Z = R$ $Z = X_L$ $Z = XC$
Resistance and   Resistance and   Inductance and
Inductance only: Capacitance only: Capacitance only:
$Z = \sqrt{R^2 + X_L^2}$ $Z = \sqrt{R^2 + X_C^2}$ $Z = \sqrt{X_L - X_C)}$

Ohm's law for AC circuits:

$E = 1 \times Z$ $I = \dfrac{E}{Z}$ $Z = \dfrac{E}{I}$

Capacitances in Parallel:
$C_t = C_1 + C_2 \; C_3 + ...$
C_t = total capacitance (farads)
$C_1 C_2 C_3 ...$ = individual cpacitances (farads)

Capacitances in series:

$$C_t = \dfrac{1}{\dfrac{1}{C_1} + \dfrac{1}{C_2} + \dfrac{1}{C_3} + ...}$$

Phase Angle

An alternating current through an inductance lags the voltage across the inductance by an angle calculated as follows:

Tangent of angle of lag = $\dfrac{X_L}{R}$

An alternating current through a capacitance leads the voltage across the capacitance by an angle calculated as follows:

Tangent of angle of lead = $\dfrac{X_C}{R}$

The resultant angle by which a current leads or lags the voltage in an entire circuit is called the phase angle and is calculated as follows:

Cosine of phase angle = $\dfrac{R \text{ of circuit}}{Z \text{ of circuit}}$

Power Factor

Power factor of a circuit or system is the ratio of actual power (watts) to apparent power (volt-amperes), and is equal to the cosine of the phase angle of the circuit:

$PF = \dfrac{\text{actual power}}{\text{apparent power}} = \dfrac{\text{watts}}{\text{volts} \times \text{amperes}} = \dfrac{kW}{kVA} = \dfrac{R}{Z}$

KW = kilowatts

kVA = kilowatt-amperes = volt-amperes + 1,000

PF = power factor (expressed as decimal or percent)

Single-Phase Circuits

$kVA = \dfrac{EI}{1,000} = \dfrac{kW}{PF} \quad kW = kVA \times PF$

$I = \dfrac{P}{E \times PF} \quad E = \dfrac{P}{I \times PF} \quad PF = \dfrac{P}{E \times I}$

$P = E \times I \times PF$

P = power (watts)

Two-Phase Circuits

$I = \dfrac{P}{2 \times E \times PF} \quad E = \dfrac{P}{2 \times I \times PF} \quad PF = \dfrac{P}{E \times I}$

$kVA = \dfrac{2 \times E \times I}{1000} = \dfrac{kW}{PF} \quad kW = kVA \times PF$

$P = 2 \times E \times 1 \times PF$

E = phase voltage (volts

Three-Phase Circuits, Balanced Star or Wye

$I_N = 0 \quad I = I_P \quad E = \sqrt{3} E_P = 1.73 E_P$

$E_P = \dfrac{E}{\sqrt{3}} = \dfrac{E}{1.73} = 0.577 E$

I_N = current in neutral (amperes)

I = line current per phase (amperes)

I_P = current in each phase winding (amperes)

E = voltage, phase to phase (volts)

E_P = voltage, phase to neutral (volts)

Three-Phase Circuits, Balanced Delta

$I = 1.732 \times I_P \quad I_P = \dfrac{1}{\sqrt{3}} = 0.577 \times I$

$E = E_P$

Power:

Balanced 3-Wire, 3-Phase Circuit, Delta or Wye

For unit power factor (PF = 1.0):

$P = 1.732 \times E \times I$

$I = \dfrac{P}{\sqrt{3}} E = 0.577 \dfrac{P}{E} \qquad E = \dfrac{P}{\sqrt{3}} \times I = 0.577 \dfrac{P}{I}$

P = total power (watts)

For any load:

$P = 1.732 \times E \times I \times PF \quad VA = 1.732 \times E \times I$

$E = \dfrac{P}{PF \times 1.73 \times I} = 0.577 \times \dfrac{P}{PF \times I}$

$I = \dfrac{P}{PF \times 1.73 \times E} = 0.577 \times \dfrac{P}{I} \times E$

$PF = \dfrac{P}{1.73 \times I \times E} = \dfrac{0.577 \times P}{I \times E}$

VA = apparent power (volt-amperes)

P = actual power (watts)

E = line voltage (volts)

I = line current (amperes)

Power Loss:

Any ac or dc Circuit

$P = I_2 R I = \sqrt{\dfrac{P}{R}} R = \dfrac{P}{I^2}$

P = power heat loss in circuit (watts)

I = effective current in conductor (amperes)

R = conductor resistance (ohms)

Load Calculations

Branch Circuits—Lighting & Appliance 2-Wire:

$$I = \frac{\text{total connected load (watts)}}{\text{line voltage (volts)}}$$

I = current load on conductor (amperes)

3-Wire:

Apply same formula as for 2–wire branch circuit, considering each line to neutral separately. Use line-to-neutral voltage; result gives current in line conductors.

USEFUL FORMULAS

TO FIND	SINGLE PHASE	THREE PHASE	DIRECT CURRENT
AMPERES when kVA is known	$\dfrac{\text{kVA} \times 1000}{E}$	$\dfrac{\text{kVA} \times 1000}{E \times 1.73}$	not applicable
AMPERES when horsepower is known	$\dfrac{\text{HP} \times 746}{E \times \% \text{ eff.} \times \text{pf}}$	$\dfrac{\text{HP} \times 746}{E \times 1.73 \times \% \text{ eff.} \times \text{pf}}$	$\dfrac{\text{HP} \times 746}{E \times \% \text{ eff.}}$
AMPERES when kilowatts are known	$\dfrac{\text{kW} \times 1000}{E \times \text{pf}}$	$\dfrac{\text{kW} \times 1000}{E \times 1.73 \times \text{pf}}$	$\dfrac{\text{kW} \times 1000}{E}$
HORSEPOWER	$\dfrac{I \times E \times \% \text{ eff.} \times \text{pf}}{746}$	$\dfrac{I \times E\ 1.73 \times \% \text{ eff.} \times \text{pf}}{746}$	$\dfrac{I \times E \times \% \text{ eff.}}{746}$
KILOVOLT AMPERES	$\dfrac{I \times E}{1000}$	$\dfrac{I \times E \times 1.73}{1000}$	not applicable
KILOWATTS	$\dfrac{I \times E \times \text{pf}}{1000}$	$\dfrac{I \times E \times 1.73 \times \text{pf}}{1000}$	$\dfrac{I \times E}{1000}$
WATTS	$E \times I \times \text{pf}$	$E \times I \times 1.73 \times \text{pf}$	$E \times I$

$$\text{ENERGY EFFICIENCY} = \frac{\text{Load Horsepower} \times 746}{\text{Load Input kVA} \times 1000}$$

$$\text{POWER FACTOR (pf)} = \frac{\text{Power Consumed}}{\text{Apparent Power}} = \frac{W}{VA} = \frac{\text{kW}}{\text{kVA}} = \cos\varnothing$$

I = Amperes
HP = Horsepower

E = Volts
% eff. = Percent Efficiency
e.g., 90% eff. is 0.90

kW = Kilowatts

kVA = Kilovolt-amperes
pf = Power Factor
e.g., 95% pf is 0.95

EQUATIONS BASED ON OHM'S LAW:

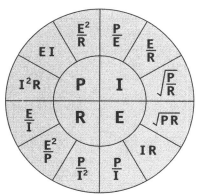

P = POWER, IN WATTS
I = CURRENT, IN AMPERES
R = RESISTANCE, IN OHMS
E = ELECTROMOTIVE FORCE, IN VOLTS

Appendix 3: Answer Key

SAMPLE EXAM QUESTION REVIEW

1. A. Ref.—Article 350.10
2. B. Ref.—352.26
3. C. Ref.—360.20(B)
4. C. Ref.—314.28(A)(2)
5. B. Ref.—300.6(C)
6. B. Ref.—368.30
7. C. Ref.—Chapter 9, Table 1
8. B. Ref.—400.13 and 240.5(A)
9. D. Ref.—310.14
10. D. Ref.—300.4(F)
11. D. Ref.—334.30
12. A. Ref.—Table 310.15(B)(6)
13. D. Ref.—300.5(D)
14. D. Ref.—Table 310.15(B)(6)
15. D. Ref.—280.25
16. C. Ref.—408.15
17. B. Ref.—230.54(C) Exception
18. A. Ref.—230.31(B) Exception
19. B. Ref.—230.9(A) Exception
20. C. Ref.—230.24(A)
21. D. Ref.—240.83(C)
22. D. Ref.—110.26(D)
23. B. Ref.—230.95
24. D. Ref.—General knowledge; Appendix 2
25. B. Ref.—Table 430.37
26. A. Ref.—430.81(B)
27. B. Ref.—404.8(B)
28. C. Ref.—450.43(A)
29. A. Ref.—300.7(A)
30. B. Ref.—410.66(A)
31. C. Ref.—410.76(B)
32. A. Ref.—700.12(B)(2)
33. C. Ref.—520.25(A) and (C)
34. D. Ref.—501.15(A)
35. A. Ref.—500.5(C)
36. D. Ref.—511.4(A)(1)
37. B. Ref.—517.18(B)
38. C. Ref.—675.15, 250.106 (FPN) 2
39. A. Ref.—440.61, 440.3, and 250.114(3)(A)
40. B. Ref.—250.66(A)
41. D. Ref.—250.52(A)
42. C. Ref.—General knowledge; Appendix 2
43. A. Ref.—210.52, 210.11(C)
44. B. Ref.—230.95(A)
45. C. Ref.—Chapter 9, Tables 1, 4, 5
46. C. Ref.—424.3(B)
47. B. Ref.—Chapter 3, this text
48. D. Ref.—90.2(B)
49. C. Ref.—Chapter 2 text
50. C. Ref.—460.2

QUESTION REVIEW CHAPTER 2

1. No. However, there is an exception: 356.12 for electrical signs. Ref.—356.12 Exception

2. Governmental bodies exercising legal jurisdiction over the electrical installation, insurance inspectors, or others with the authority and responsibility for governing the electrical installation. Ref.—90.4

3. Ships, railway rolling stock, aircraft, or automotive vehicles. Ref.—90.2(B). Four installations are not covered, for example, see 90.2(B)(1)–(5).

4. Fine print notes (FPN). Ref.—90.5(C)

5. 110.3(B)

6. When adopted by the regulatory authority over the intended use. Ref.—Pages 70-i, which states that the *Code* is advisory as far as NFDA and ANSI are concerned but is offered for use in law for regulatory purposes.

7. No. It is the intent that all premise wiring or other wiring of the utility-owned meter read equipment on the load side of service point of buildings, structures, or other premises not owned or leased by the utility are covered. It also is the intent that buildings used by the utility for purposes other than listed in 90.2(B)(5), such as office buildings, warehouses, machine shops, or recreational buildings, that are not an integral part of the generating plant, substation, or control center are also covered. Ref.—90.2(B)(5)

8. There has been a change in this edition of the *NEC®* from the wording of the previous edition.

9. (A) 2005, (B) September 2004.

10. 1881, The National Fire Engineers met in Richmond, Virginia, which resulted in the first document that led to the *National Electrical Code®*. Various meetings were held after that; then in 1897 some various allied interests of the church, electrical, and architectural concerns met, and in 1911 NFPA was set up as the sponsor.

11. Yes. Ref.—Article 590, 590.4(C)

12. Chapter 8. Ref.—90.3

13. No, it is not a training manual for untrained persons. Ref.—90.1(C)

14. No, only underground mines and rolling stock for surface mines. All surface mining other than the rolling stock are covered by the *NEC®*. Ref.—90.2(B)(2)

15. Nowhere in the *Code* does it state that it is the minimum allowed. Only when adopted by a local jurisdiction would it be considered the minimum of that jurisdiction. When an installation is installed for the *National Electrical Code®*, it provides a safeguarding of persons and property, and the provisions are considered necessary for safety and compliance with the *NEC®* will result in an installation essentially free from hazards. However, it may not be efficient or convenient or adequate in some installations. Ref.—90.1(A) and (B)

16. Definitions, Article 100, to continue for three hours or more

17. 1999. *The National Electrical Code®* is adopted every three years; there being a 1999, a 2002, and a 2005. Therefore, not adopting the two editions, they would be using the 1999 *NEC®*.

18. Article 725 covers Class 2 wiring.

19. Article 517—Health Care Facilities, 517.2, Definitions

20. 250.52(A)(2). The water pipe must be used if available, and supplemented with another grounding electrode of one of the types listed in 250.52 (A)(1) through 250.52(A)(7).

QUESTION REVIEW CHAPTER 3

(Reduce fractions to their lowest terms.)

1. ½

2. ¼

3. $\frac{1}{10}$

4. ⅖

5. ¾

6. ¾

7. ½

8. ⅗

9. ⅗

10. ⅕

11. ¼

12. ¼

(Change mixed numbers to improper fractions.)

1. $\frac{3}{2}$

2. $\frac{13}{4}$

3. $\frac{43}{8}$

4. $\frac{17}{4}$

5. $\frac{57}{8}$

6. $\frac{13}{6}$

7. $\frac{23}{4}$

8. $\frac{25}{3}$

9. $\frac{46}{7}$

10. $\frac{29}{8}$

(Change improper fractions to mixed numbers.)

1. 4½

2. 2⅖

3. 12⅘

4. 3¼

5. 9⅔

6. 15½

7. 1⅔

8. 4⅓

(Multiply whole numbers and fractions. Answer in proper reduced form.)

1. 4

2. ⅜

3. $\frac{1}{40}$

4. $\frac{9}{2}$ = 4½

5. $\frac{21}{4}$ = 5¼

6. $\frac{5}{2}$ = 2½

7. 9

8. $\frac{15}{7}$ = 2⅐

(Convert fractions to decimal equivalents.)

1. .25
2. .625
3. .75
4. .866
5. .375

(Resolve to whole numbers.)

1. 16
2. 81
3. 2500
4. 1
5. 1000

(Find the square root.)

1. 5
2. 9
3. 7
4. 56
5. 101
6. 29
7. 78

(Solve the following problems.)

1. .05 or 5%. Ref.—Ohm's law; Appendix 2
2. 100%. Ref.—Chapter 3
3. 5 amperes. Ref.—Chapter 3, Ohm's law; Appendix 2
4. 1.33 ohms. Ref.—Chapter 3
5. 14.4 ohms. Ref.—Chapter 3
6. 6 volts. Ref.—Chapter 3
7. 33.34 ohms
8. 83.34 ohms
9. 23.81 ohms
10. 107.15 ohms
11. 37 volts
12. 56 volts
13. 27 volts
14. 1.12 amperes

CHAPTER 3—PRACTICE EXAM

1. C
2. D
3. D
4. C

5. C
6. A
7. D
8. B
9. C
10. B
11. C
12. A
13. B
14. C
15. A

QUESTION REVIEW CHAPTER 4

1. Yes. Ref.—90.2(A). A floating restaurant must be wired as a typical commercial restaurant.
2. Chapter 8. Article 250 does not apply in Chapter 8 unless specifically referenced.
3. Article 406
4. False. Branch-circuits are referenced in many articles in the *NEC*®.
5. Article 516. Luminaires (fixtures) installed in a booth must be type Class I, Division 1, generally.
6. Article 422 for appliances, specifically 422.2. Electrical heating is covered in Article 424 and heat pumps in Article 440.
7. Chapter 9, Tables 1 through 8 and all Notes
8. Lighting and power in mine shafts are not covered by the *NEC*®, Article 90, 90.2(B)
9. 300.50(A)
10. Article 225, which covers outside branch-circuits and feeders, 225.31
11. Chapter 9, Table 10. Expansion fittings are required by 300.7, and 352.44, Tables 352.44(A) and 352.44(B).
12. Places of assembly, Article 518
13. *NEC*® Article 250, 250.34
14. Article 680, 680.26 Part II
15. Article 90, 90.9(C)(2)
16. These requirements appear in the building codes or the NFPA Life Safety Code; they are not found in the *NEC*®.
17. Article 760
18. Yes. Ref.—590.6(A)

19. Article 502—(a) Class 2, (b) Division 1. Article 500 for the definitions and Article 502(10)(A) and (B) for the specific requirements.

20. It is a violation to put over 1000-volt wiring in a residence. Neon signs requirements are found in Article 600. However, Article 410 requirements limit installation under 1000 volts in dwellings. Ref.—410.75 and 410.80(B)

QUESTION REVIEW CHAPTER 5

1. 4 feet (1.2 m). Ref.—110.26, Table 110.26(A). Some inspectors may permit condition 2.

2. Two. Ref.—110.26(C)

3. Yes. Ref.—110.26(D)

4. 6½ feet (2 m). Ref.—110.26(E)

5. One. Where the required working space is doubled, only one door is required. Ref.—110.26(C)

6. No. The required disconnecting means and supplementary overcurrent device must have a working clearance in accordance with 110.26. Clear working clearance of 30 inches (750 mm) wide and 40 inches (1 m) deep is required from the ground level up 6½ feet (2 m).

7. 30 inches (750 mm). Ref.—110.26(A)(2)

8. Ungrounded. Ref.—240.20(A)

9. 60 percent. Ref.—Chapter 9, Table Note (4)

10. Clothes closet. Ref.—240.24(D)

11. 1¼ inches (32 mm). Ref.—300.4(A)(1)

12. 600 volts. Ref.—362.12(6)

13. Yes. Ref.—230.43(16)

14. Yes. It shall be bonded to the electrode at each end and all intervening raceway boxes and enclosures between the service equipment and the grounding-electrode conductor. Ref.—250.92(A)(3)

15. Overcurrent device. Ref.—250.122

16. Two. Ref.—210.52(E)

17. No. Ref.—680.74, except manufacturer's instructions may require 110.3(B)

18. Yes. Ref.—210.52(A)

19. No. Ref.—210.52(A) for site-built dwellings; 550.13(D) Exception 4 for mobile homes

20. 210.63. Ref.—Article 210

QUESTION REVIEW CHAPTER 6

1. 10, 80 percent loading. Ref.—220.11

2. 1120 square feet. Ref.—Knowledge; 220.12

3. 2. One additional 20-ampere circuit must be added for the bathroom receptacles. Ref.—Table 220.12 and Footnotes, also see Figure 6–4, 210.11(C)(2)

4. 5—2 kitchen appliance circuits, 1 laundry circuit, 1 bathroom receptacle, and 1 heating circuit. Ref.—422.12, 210.11(C)

5. 2. Ref.—210.11(C)(1)

6. Yes. Ref.—210.11(C)(2)

7. 125 percent of nameplate. Ref.—422.13 and 422.10

8. None. The *NEC*® does not require power to the structure. Ref.—210.52(G)

9. 2. Ref.—210.52(E)

10. Four minimum, one for the required bathroom circuit, one for the required outdoor receptacles, and two for the required kitchen appliance circuit minimum. Ref.—210.8(A), also see 210.52 and all subsections

11. Yes. Illumination is required. Ref.—210.70(A)(2)(b)

12. No. Ref.—210.70

13. Yes. Ref.—210.63

14. 20 ampere. Ref.—210.11(C)(2)

15. No. Ref.—422.12, Definition of Individual Circuit (Article 100)

16. No. Ref.—410.8

17. Yes. Ref.—210.52(B)(2) Exception 2

18. Yes. Ref.—Definition of direct grade access. Ref.—210.8(3) and 210.52(E).

19. Yes, two receptacles at direct grade access are required. Ref.—210.52(E) and 210.8

20. 230.79(C)

    (a) 220.12(J) and Table 220.12; $28 \times 40 = 1120$ sq. ft.; $3 \text{ VA} \times 1120 = 3360$ volt-amperes

    (b) 220.52; 3000 volt-amperes

    (c) 220.14(A); 4500 volt-amperes, 422.13

    (d) 220.52; 1500 volt-amperes

    (e) 220.54; 5000 volt-amperes

    (f) 0. Electricity not required; however, if supplied, the load would have to be included in calculation. Ref.—210.52(G)

(g) 220.14(A); size not stated, therefore, a 20-ampere branch-circuit is recommended.

(h) 0. 220.14(A); 210.52(B) Example 2 could be general lighting or on appliance load

(i) 0. None. Read question again.

(j) 0. Gas range igniter can be fed from appliance circuit, 210.52(B)(2) Exception 2

(k) 0. None. This load is included in general lighting calculations, 220.3.

QUESTION REVIEW CHAPTER 7

1. 6. Ref.—230.208

2. A disconnecting means is required at each building. Ref.—Article 225 Part II

3. Yes, you are limited to not more than six disconnects on this secondary side. Ref.—450.3(B)—conductors must be protected in accordance with 240.4.

4. No; however, 230.54(E) requires that the conductors be separated where of a different potential.

5. No. 240.24(E) prohibits branch-circuit overcurrent devices from being located within a residential bathroom. Ref.—240.24(E), 230.70(A)(2)

6. Yes. SE cable can be used in the same manner as Type NM cable. However, it is SE cable, and must comply with Articles 300 and 334, Part II. Ref.—338.10(B)(4)

7. Usually no. Yes, if the window is not designed to be opened, or where the conductors run above the top level of the window. Where windows are designed for opening and closing, then a minimum of 3 feet (900 mm) from the windows, doors, porches, fire escapes, or similar locations is required. Ref.—230.9

8. (a) Yes. Ref.—250.64(D)

   (b) They shall be sized in accordance with 250.66 for the largest conductor entering each respective enclosure.

9. Requirement to open all of the conductors is a disconnect requirement and not an overcurrent requirement. Ref.—240.20(B) gives the requirements for circuit breakers, 240.40 gives the requirements for fuses. 210.4 is the requirement for multiwire branch-circuits. Disconnecting, not overcurrent, is the purpose.

10. A separately derived system is defined in Article 100, and the grounding installation requirements are found in 250.30. Ref.—Article 100, 250.30

11. 450.3—not more than six disconnects are permitted on the secondary side of the transformer.

12. No. Identification of the hi-leg is only required where the neutral is present. Ref.—110.15, 230.56, and 408.3(E)

13. No. The grounding-electrode conductor must be continuous, and the connection shall be made solidly and used for no other purpose. Ref.—250.64(C)

14. Less than ¼ inch (6.4 mm). Ref.—352.44 Exception

15. Grounded effectively is defined in Article 100. The authority having jurisdiction will make that determination.

16. Article 334, nonmetallic-sheathed cable. Romex is a trade name.

17. Article 360 covers the requirements of flexible nonmetallic tubing. It is manufactured in trade sizes ½ (16) and ¾ (21) only. *Note:* This is a flexible metallic tubing that is liquidtight without a nonmetallic jacket; rarely used in construction installation.

18. Article 404

19. Chapter 9, Table 1, Footnote 9 states that multiconductor cable of two or more conductors shall be treated as a single conductor cable, and cables that have an elliptical cross-section, the cross-section layering calculation shall be based on using the major diameter for the ellipse as the circle diameter.

20. Article 511 does not apply. 511.1 and 511.2 clearly exempt these parking structures. Therefore, Chapters 1 through 4 apply, and the wiring method would be taken from those four chapters.

QUESTION REVIEW CHAPTER 8

1. Yes. Ref.—240.16(A). However, handle ties are required.

2. No maximum length needed, provided they are outdoors. Ref.—240.21(B)(5)

3. No. Ref.—240.60(A). Voltage to ground can exceed 300 volts on a wye connection and, therefore, would not be permitted.

4. It is not a limiting factor. However, the six disconnect rule applies. Ref.—230.90(A) Exception 3

5. 100 percent. Ref.—424.3 and 220.51

6. No. Ref.—501.10(B), 350.60, 501.30(B), and must comply with 250.102

7. Yes. Both the overcurrent device and conductors would have to meet these requirements. These are Exceptions to the general rule. Ref.—215.2(A)(1)

8. There is no limit. Ref.—Table 220.12 and 220.14(3). *Note:* Many local jurisdictions set limits on 15- and 20-ampere branch-circuits in all installations. However, the *NEC®* does not specify a limit on dwelling branch-circuits for general-use electrical outlets.

9. No. Ref.—220.14(I), also 240.4(D) limits 12 AWG conductors to a 20-ampere branch-circuit. 16×180 VA $= 2880 \div 120$ V $= 24$ amperes.

10. (A) No. Ref.—220.60

    (B) Yes. Ref.—424.3(B)

11. No, only if installed in cable with 12 inches (300 mm) of cover. Otherwise, can be installed at the greater depth or in raceways. Ref.—300.5 and Table 300.5

12. Yes. Ref.—240.16(A)

13. No. Ref.—215.2. Footnote: Branch-circuits are required to be calculated at 125 percent for water heaters—422.13.

14. Yes. Ref.—210.19 and 110.3(B)

15. No. Ref.—210.6—12 amperes or less must be 120 volts; over that can be 240 volts. 1400 VA $\div$ 240 V $= 5.833$ amperes

16. 180 VA. Ref.—220.14(I)

17. Yes, it complies with the tap rules in 240.21(B) for the 10-foot tap rules.

18. Yes, it complies with the 25-foot tap rule. Ref.—240.21(B)(2)

19. No, it does not comply with any of the tap rules in 240.21(B)(2)

20. True. Ref.—368.17(B)

QUESTION REVIEW CHAPTER 9

1. Soldered. Ref.—250.8

2. 14. Ref.—600.7(D)

3. Not required to be accessible. Ref.—250.68(A)

4. White or gray. Ref.—200.6(A)

5. 4/0. Ref.—250.24

   3-500 kcmil = 1500 kcmil or 3 each 500 kcmil = 1500 kcmil $\times$ 12½% = 187.5 kcmil

   $1500 \times 12.5\% = 187.5$ kcmil

   Chapter 9, Table 8—3/0 AWG 167.8 kcmil too small; therefore, use 4/0 AWG 211.6 kcmil

6. 4. Ref.—Table 250.102(D), 250.122

7. Yes, bonding is not required. Ref.—250.94(3)

8. They are required to be run in each nonmetallic raceway and are required to be sized in accordance with the sum of the paralleled phase conductors. Ref.—215.6 and 250.122 and Table 250.122

9. Yes. Ref.—250.112(L)

10. No, the language is specific. Ref.—250.30

11. Yes, it must be bonded. No, it is not sufficient to bond at one end. It must be bonded at both ends and at all enclosures in which it passes through. Ref.—250.64(E)

12. No, a driven rod is not acceptable. An equipment grounding conductor must be carried with the circuit. Ref.—551.76(A)

13. Yes, the requirements for separately derived transformers are found in 250.30. These requirements do not amend the requirements of 250.50, but they are the specific requirements. Therefore, a supplementary grounding electrode is required.

14. 8 AWG copper or equivalent. Ref.—551.56(C)

15. No. Ref.—250.104. *Note:* The interior gas piping system is required to be grounded for safety. It can be grounded with the circuit conductor equipment ground most likely in which it would come in contact. For example, a central heating electric furnace may have 12 AWG conductors supplying the fan motor. The 12 AWG equipment grounding conductor run with those circuit conductors could be the conductor used to bond the gas piping system in this case. **Warning:** The underground gas piping system may not be grounded. Electrolysis may occur, creating dangerous underground gas leaks.

16. Yes, a 1 AWG copper or a 2/0 AWG aluminum or copper-clad aluminum wire is required in each conduit. Ref.—Table 250.122 and 250.122, which state that where run in parallel in multiple raceways or cables as permitted, the equipment grounding conductors also shall be run in parallel,

each equipment grounding conductor shall be sized on the basis of the ampere rating of the over-current device protecting the circuit conductor in the raceway in accordance with Table 250.122.

17. No. Ref.—110.14(B)

18. 8 AWG. Ref.—Table 250.122. This is a minimum size conductor and may have to be larger if the fault current available exceeds that for which the 8 AWG conductor will carry.

19. Yes. The flexible metal conduit cannot exceed 6 feet (1.8 m) for the combined two pieces of flexible metal conduit. Ref.—250.118, 348.60

20. The maximum rating of the grounding impedance, no smaller than 8 AWG copper or 6 AWG aluminum. Ref.—250.24, 250.36

QUESTION REVIEW CHAPTER 10

1. Within 12 inches (300 mm) of the outlet or box, and every 4½ feet (1.4 m). Ref.—338.10(4) and 334.30. The 2005 *NEC*® requires a separate equipment grounding conductor for all new installations. Therefore, Type SE cable may no longer be appropriate, Type SER would be acceptable.

2. Maximum number of conductors shall be permitted using the provisions of 314.16(B). Conduit bodies shall be supported in a rigid and secure manner. Ref.—314.16(C)

3. Yes. Ref.—348.20(A)(1) and 350.20(A) Exception

4. Yes. Ref.—342.30(A)

5. Yes. Ref.—362.30(B)—Metal studs are the required support when they do not exceed 3 feet (900 mm) apart.

6. Type NM cable shall be supported at intervals not exceeding 4½ feet (1.4 m) and within 12 inches (300 mm) of every cabinet, box, or fitting. Two conductor cables shall not be stapled on edge. Where Type NM cable is run through holes in wood studs or rafters it shall be considered supported. Ref.—334.30

7. No. Ref.—378.12

8. Aboveground conductors must be installed in a rigid metal conduit, intermediate metal conduit, electrical metallic tubing, rigid nonmetallic conduit in cable tray or busway or cable bus or in other identified raceways, or as open runs of metal clad cable suitable for the use. Open runs of Type MV cable, bare conductors, or bare bus bars also are permitted as they apply. Ref.—300.37

9. Yes, provided that carpet squares not exceeding 36 inches square (914 mm²) are used in a wiring system in accordance with Article 324 Type FCC cable and associated accessories are used. Ref.—Article 324.10(H)

10. Metallic wireway shall be established in accordance with 376.22 and shall not exceed 30 current-carrying conductors at any cross-sectional area. The sum of cross-sectional areas of all conductors at any cross-section of the wireway shall not exceed 20 percent. The derating factors in Table 310.15(B)(2) *do not* apply. For nonmetallic wireway, 378.22 applies. The cross-sectional area containing conductors at any cross-section of the nonmetallic wireway shall not exceed 20 percent of the interior cross-sectional area. The derating factors in Table 310.15(B)(2) *do* apply to the current- carrying conductors at the 20 percent specified. *Note:* This is a critical difference. Caution should be used when reading questions as they apply to a specific wiring method, metallic or nonmetallic, because the conditions and rules are very different.

11. 1¼ inches (32 mm) from the nearest edge of the stud. Ref.—300.4(D)

12. There is no difference in the permitted uses or installation requirements in the two products. Article 342 and Article 344 are essentially the same. Ref.—Article 342 and Article 344

13. 1/0 AWG. Ref.—310.14

14. Yes, in any building exceeding three floors, nonmetallic tubing must be concealed within a wall, floor, or ceiling where the material has at least a 15-minute finished rating as identified in the listings of the fire-rated assemblies. *Exception: Where a fire sprinkler system(s) is installed in accordance with NFPA 13-1999,* Standard for the Installation of Sprinkler Systems, *on all floors, ENT is permitted to be used within walls, floors, and ceilings, exposed or concealed, in buildings exceeding three floors above grade* [reprinted with permission from NFPA 70-2005]. Ref.—362.10(2)

15. No, they must be removed. They are not permitted for a period to exceed 90 days. Ref.—590.3(B)

16. No, lengths not exceeding 4 feet (1.2 m) can be used to connect physically adjustable equipment

and devices that are permitted in the duct or plenum chamber. Ref.—300.22(B)

17. ¼ inch (6 mm). Ref.—352.44

18. Yes, even though 300.3 requires that all conductors of the same circuit, including the neutral and equipment grounding conductors, must be run within the same raceway, cable trench, cable, or cord; Exception 2 to that section permits that with column-type panelboards; the auxiliary gutter and pull box may contain the neutral terminations. Ref.—300.3(B)(4)

19. No. Generally, each cable shall be secured to the panelboard individually. Ref.—312.5(C). The Exception would allow this under conditions.

20. No, a branch-circuit or a lighting panelboard is one that 10 percent of its overcurrent devices are rated 30 amperes or less for which neutral connections are provided. A lighting and branch-circuit panelboard, as required by the *Code*, shall have not more than two circuit breakers or two sets of fuses for disconnecting. Ref.—408.14 and 408.16—A disconnect is required ahead of the lighting and branch-circuit panelboard although the load is relatively small in this cabin.

QUESTION REVIEW CHAPTER 11

1. 430.6(A). Ref.—430.6(A)

2. 100 ampere. Ref.—430.6, Tables 430.250 and 430.52 (40 × 250% = 100)

3. 175 percent. Ref.—Tables 430.250 and 430.52

4. 14 AWG. Ref.—Table 210.24

5. 20. Ref.—240.4, 240.6, and Table 310.16

6. 40. Ref.—240.4, 240.6, and Table 310.16

7. 100. Ref.—240.4, 240.6, and Table 310.16

8. 200. Ref.—240.4, 240.6, and Table 310.16

9. 125 percent. Ref.—450.3(B)

10. 175 percent. Ref.—Table 430.52

11. 125 percent. Ref.—Article 430.32(A)(1)

12. 6. Ref.—Table 430.91

13. Indoors. Ref.—410.73(E) and (F)

14. Fire protection or equipment cooling. Ref.—450.47

15. Accessible. Ref.—450.13(A)

16. 3. Ref.—460.2

17. 8. Ref.—440.32 and Table 310.16

18. 112.5 kVA. Ref.—450.21(B)

19. 600 watts. Ref.—Ohm's law (*Clue:* Resistance does not change); Appendix 2

20. 87 percent. Ref.—Chapter 3; Appendix 2

QUESTION REVIEW CHAPTER 12

1. Nonconductive, conductive, composite. Ref.—770.5

2. True. Ref.—511.7

3. Yes, where the raceway emerges from the ground, both at the office and at the dispenser. Ref.—514.9(A) and (B)

4. False. Ref.—700.1

5. Class 1, Division 2. Ref.—511.3(B)(2)

6. Inadvertent parallel interconnection. Ref.—700.6

7. ⅝ (18 mm). Ref.—501.15(C)(3)

8. 12 inches (300 mm). Ref.—490.24

9. Two duplex or four single receptacles. Ref.—517.18(B)

10. 6. Ref.—220.14 and 430.24, Tables 430.248 and 310.16

11. 15 feet (4.5 m). Ref.—600.10(D)(2)

12. False. Ref.—240.53

13. True. Ref.—240.53

14. False. Ref.—240.54

15. True. Ref.—Bussmann SPD Protection Handbook, 240.53(B)

16. (a) Electrician or electrical contractor

    (b) Field marking is required to state "Caution—Series Rated System, ___ Amperes Available. Identified Replacement Components Required." Ref.—110.22

    (c) Manufacturer

17. 6 amperes. Ref.—$I = \dfrac{HP \times 746}{E \times PF \times \text{eff}\%}$

    Note: When power factor is not noted, it must be calculated at 100 percent.

18 60. Ref.—Tables 430.250 and 430.52

19. 27. Ref.—Ohm's law, Chapter 3, Table 220.20; Appendix 2

20. 18. Ref.—Ohm's law, Appendix 2

QUESTION REVIEW CHAPTER 13

1. Yes. Grounding electrode. Ref.—810.20(C), 810.21

2. No. Chapter 8 stands alone. There are no burial depth requirements for satellite dishes installed anywhere there is no safety hazard, and, therefore, the only consideration is for the customer's protection of his or her own equipment, *NEC*® 800.10. *Note:* Also look at 90.3. Chapter 8 covers communication systems and is independent of the other chapters, except where they are specifically referenced. Chapter 8 stands alone, and other chapters cannot be applied unless specifically referenced.

3. Yes. Ref.—800.33

4. 6 feet (1.8 m). Ref.—800.13

5. It is an intrinsically safe circuit covered in Article 504 of the *National Electrical Code*®.

6. Article 406 covers these isolated ground receptacles intended for the reduction of electrical noise. The *NEC*® requires receptacles used for "isolated ground" be marked on the face of the receptacle with an orange triangle. The solid orange receptacle is no longer acceptable. Ref.—Article 406 and 250.96(B)

7. Yes. Ref.—Article 650

8. Yes, Article 110, 110.26 for installations of less than 600 volts, 110.32 for installations over 600 volts.

9. Yes. 110.12(C) clearly states that paint, plaster, and other abrasives may damage the bus bars, wiring terminals, insulators, and other surfaces in panelboards. Therefore, the internal parts of this equipment must be covered and protected.

10. Inspector was correct. 550.32 requires the service equipment to be located outside and not more than 30 feet (9 m) from the exterior wall of the mobile home. An exception permits the installation of the service equipment on the mobile home when installed by the manufacturer. Electrical contractors, however, cannot make this installation.

11. Optional standby system is covered in Article 702 and is defined in 702.2.

12. Color coding is not specifically required; however, 210.4(D) requires the identification of ungrounded conductors on multiwire branch-circuits under certain conditions. This identification can be by color coding, marking tape tagging, or other equally effective means.

13. 225.7(B) covers outdoor branch-circuits for lighting equipment and there is no maximum number. See 215.4(A) for feeders.

14. Yes. Article 547—Agriculture Buildings covers this type of installation.

15. Yes. Article 324.10(H) covers this type of installation. However, the carpet must be laid in squares not to exceed 3 feet × 3 feet (36 inches [914 mm] square).

16. Article 280—Surge Arresters covers that type of installation.

17. Article 590 permits a class less than would be required for permanent installation. However, there are time constraints. *Note:* 590.4(A) requires the service on that temporary pole to comply with Article 230.

18. 86 percent. Ref.—Chapter 3, Duff and Herman, *Alternating Current Fundamentals, 4E* by Thomson Delmar Learning

19. 60 hp. EFF = $\dfrac{W}{\sqrt{3} \times E \times I}$; Appendix 2

Ref.—Loper, *Direct Current Fundamentals, 4E* by Thomson Delmar Learning; Appendix 2

20. 0.105. Ref.—Loper, *Direct Current Fundamentals, 4E* by Thomson Delmar Learning; Appendix 2

PRACTICE EXAMINATION

1. C. Ref.—Article 100—Definitions

2. A. Ref.—310.11(C)

3. E. Ref.—725.23

4. D. Ref.—312.8

5. B. Ref.—240.51(A) and 240.53(A)

6. D. Ref.—760.52(A)

7. C. Ref.—Chapter 9, Note 4 to Table 1

8. D. Ref.—240.80

9. C. Ref.—720.5

10. D. Ref.—Table 402.5

11. A. Ref.—Chapter 9, Table 8, Col 6

12. B. Ref.—760.54, 55, and 58

13. B. Ref.—90.5(C)

14. D. Ref.—344.30 and Table 344.30(B)(2)

15. D. Ref.—Article 100, see Definition Switch

16. D. Ref.—725.11(B)

17. B. Ref.—250.12

18. C. Ref.—Article 100, definition

19. D. Ref.—110.27

20. B. Ref.—110.6

21. B. Ref.—90.9(C)(2)

22. B. Ref.—Article 100, definition

23. B. Ref.—725.54 and 725.55

24. B. Ref.—Article 100, Definition

25. A. Ref.—200.6

26. B. Ref.—110.19

27. A. Ref.—424.39

28. C. Ref.—660.5

29. A. Ref.—225.14(C)

30. C or the height of the equipment, whichever is greater. Ref.—110.26(E)

31. A. Ref.—220.14(H)

32. A. Ref.—810.11 Exception

33. B. Ref.—230.79(C)

34. A. Ref.—430.89

35. D. Ref.—225.4

36. D. Ref.—110.53

37. D. Ref.—225.36 Exception

38. A. Ref.—210.23(A)(2)

39. D. Ref.—366.23, copper (assumed) is 1000 A per square inch. 4 × ½ = 2 square inches, 1000 × 2 = 2000 amperes maximum.

40. D. Ref.—590.7

41. D. Ref.—430.32(D)(1)

42. D. Ref.—550.16(A)

43. B. Ref.—600.9

44. D. Ref.—820.40(A)(1, 3, 6)

45. E. Ref.—110.32

46. D. Ref.—398.30(E)

47. A. Ref.—Article 100, definition

48. A. Ref.—225.6(A)(1)

49. B. Ref.—210.52(B)(2)

50. B. Ref.—210.52(C)(1) through (3)

2005 *CODE* CHANGES QUIZ 1

1. Article 682—Natural and Artificially Made Bodies of Water. Ref.—Table of Contents

2. Article 353—High Density Polyethylene Conduit: Type HDPE Conduit. Ref.—Table of Contents

3. Article 590—Temporary Installations. Ref.—Index

4. Bathroom. Ref.—Article 100—Definitions

5. Unclassified locations. Ref.—500.2

6. No. Ref.—Article 80 was relocated as a new addendum G.

7. The system bonding jumper is a new term to describe the jumper used to bond the grounded conductor, grounding electrode conductor, and equipment grounds together Ref.—250.30(A)(1)

8. Grounding electrode conductor. Ref.—Article 100—Definitions

9. Handhole enclosure. Ref.—Article 100—Definitions

10. Commercial or Institutional Kitchen. Ref.—210.8(B)

11. Ground fault protection Ref.—210.(B)(4)

12. Chapter 9, Table 2. Ref.—Same

13. 90.2(B)(4). Ref.—830.1 FPN No. 2

14. The fill restrictions do not apply; therefore, there are no maximum requirements. Ref.—820.48 Exception

15. False. Ref.—513.12. The GFCI requirements only apply to receptacles used for diagnostic equipment, electrical hand tools, or portable lighting equipment.

2005 *CODE* CHANGES QUIZ 2

1. Metal raceways. Ref.—645.5(D)(6)

2. Enclosed in raceways. Ref.—610.11(A) and (B)

3. Rigid metal conduit; intermediate metal conduit. Ref.—550.15(H)

4. 230; 525.12(A) and (B). Ref.—Same

5. 506.15(A)(1) through (5) Ref.—Same

6. Resistors and reactors that are part of other apparatus are not covered by Article 470 but are generally covered by the listing for the apparatus. Ref.—470.1 Exception

7. 115. Ref.—430.110

8. Bathtubs; shower stalls. Ref.—406.8(C)

9. 410.15(B)(6); 300.19. Ref.—Same

10. 400.4 and Table 400.4. Ref.—Same

11. 600 volts. Ref.—398.12

12. No. Ref.—Article 374, Section 374.17

13. False. Ref.—392.3(E). It is not limited but can be installed in accordance with 392.3.

14. Corrosive vapors; physical damage. Ref.—368.12(A)

15. 100 feet (30.5 m) of trade size 1 (27) intermediate metal conduit (IMC). Ref.—300.1(C), Article 342, Sections 342.20 and 342.130

2005 *CODE* CHANGES QUIZ 3

1. Above ground. Ref.—326.12(3)

2. Tamper resistant type. Ref.—517–18.

3. Earth. Ref.—250.4(A)(5)

4. NFPA 496. Ref.—500.2 (FPN) under the definition of "purged and pressurized."

5. 225.6. Ref.—*300.3 Exception: Individual conductors shall be permitted where installed as separate overhead conductors in accordance with 225.6* [reprinted with permission from NFPA 70-2005].

6. Branch-circuit panelboard. Ref.—200.6(D). They must be marked with one of the prescribed methods in 200.7.

7. 8. Ref.—400.7(C)

8. Bathtub; shower space; face-up; above. Ref.—550.15(F)

9. ½ (16); one. Ref.—600.32(A)(2) and (3)

10. 630. Ref.—Same

11. The disconnecting switch must be in-site. Ref.—430.102

12. 250.30(A)(5), 250.28(A) through (D) size in accordance with Table 250.66

13. No, they must be approved. Ref.—300.5(K); definition "approved" in Article 100

14. The *Code* requires listed or marked and listed. Ref.—310.8(D), 2005 adding sleeve to I.D. listed for "sunlight resistant"

15. No, the *Code* limits are 120 volts between conductors. Ref.—210.6(A)(1)

16. Yes. Ref.—210.8(A)(3)

17. 103 is the metric designator for trade size 4 conduit and tubing. Ref.—300.1(C)

18. No, EMT is an acceptable equipment grounding conductor. Ref.—250.118

19. Article 409, Part II. Ref.—Same

20. Ungrounded. Ref.—285.4

2005 *CODE* CHANGES QUIZ 4

1. Yes. Ref.—517.13(B) Exception 2

2. Nothing. See Definition, Article 100; luminaire is an international term

3. Yes. Ref.—*NEC*® Annex G

4. Yes. 90.1(D) was added, and the term "International" appears on the cover.

5. Signaling and communications was added to eliminate any confusion that the *NEC*® does include requirements for these systems and that they are covered.

6. Yes. *NEC*® Annex A and referenced in 90.7, FPN 3

7. No, the *Code* now uses a dual measuring system; either can be used as applicable. Ref.—90.9(B)

8. Article 406 was added to contain these requirements, which were formerly found in Article 410 for the 2002 *NEC*®.

9. Yes, the reference to types of fuel was deleted, and the definition now covers all self-propelled vehicles.

10. No. Each disconnect must be marked to indicate its purpose, unless the purpose is evident.

11. Yes. 110.26(C)(2) requires panic hardware on doors to electrical equipment spaces that contain equipment rated 1200 amperes or more.

12. 410.14(A) and 410.14(B)

13. Lighting fixtures. Article 100, definition of luminaire

14. Yes. Ref.—406.4(E)

15. Yes. Ref.—210.12(B)

16. No. Ref.—210.52(C)(5)

17. Yes. 424.44(G) requires GFCI protection in these locations.

18. Yes. Ref.—200.6

19. No. 210.6(E) only permits where conditions of maintenance and supervision ensure that only qualified persons service the installation.

20. Yes. 230.22 Exception permits the grounded conductor of a multiconductor cable to be bare.

2005 *CODE* CHANGES QUIZ 5

1. Yes. Ref.—230.33 in accordance with 110.14, 300.5(E), 300.13, and 300.15

2. No. 230.44 permits cable tray as a support system. Ref.—392.2 and 392.3

3. Yes. Ref.—110.14(B)

4. No. 230.70(A)(3) permits a shunt trip device, but the disconnecting means is still required 230.66, 230.70, 230.70(A)

5. Yes. Ref.—550.16(C)(3)

6. Yes. Ref.—430.83(A)(3)

7. No. The service disconnecting means shall be installed at a readily accessible location either outside of a building or structure or inside nearest the point of entrance of the service conductors.

8. Article 647, 647.1

9. 230.82(5)

10. The overcurrent protective device(s) shall be selected or set to carry indefinitely the sum of the locked-rotor current of the fire pump motor(s) and the pressure maintenance pump motor(s) and the full-load current of the associated fire pump accessory equipment when connected to this power supply. Ref.—695.4(B)(1)

11. Yes. Ref.—210.8(B)(3)

12. 240.24(B). However, Exception 2 of that section permits them to be accessible to only authorized management personnel.

13. Yes. No circuit breakers used as switches in 120-volt and 277-volt fluorescent lighting circuits shall be listed and shall be marked SWD or HID. Circuit breakers used as switches in high-intensity discharge lighting circuits shall be listed and shall be marked as HID. Ref.—240.81 and 240.83, 404.11

14. Yes, unless they are in metal raceways. Ref.—645.5(D)(6), 725.2, 760.2, 760.3(A), 770.2, and 770.3(A)

15. Article 590, 590.3(B)

16. Yes. Ref.—250.53(D)(2)

17. Yes. 250.54 but the earth shall not be the sole equipment grounding conductor.

18. A surge arrester is a protective device for limiting surge voltages by discharging or bypassing surge current, and it also prevents continued flow of follow current while remaining capable of repeating these functions. A Transient Voltage Surge Suppressor (TVSS) is a protective device for limiting transient voltages by diverting or limiting surge current; it also prevents continued flow of follow current while remaining capable of repeating these functions. Ref.—280.2 and 285.2

19. No. Ref.—314.20, first paragraph

20. 50 pounds (23 kg). Ref.—314.27(B)

2005 *CODE* CHANGES QUIZ 6

1. 6 feet (1.8 m). Ref.—320.30(B)(3)

2. Yes. Ref.—300.7(A)

3. No. Ref.—300.11(C)

4. No. Ref.—300.22(A), (B), and (C)

5. Article 590

6. It is a device intended to provide protection from the effects of arc faults by recognizing characteristics unique to arcing and by functioning to deenergize the circuit when an arc fault is detected. Ref.—210.12

7. No. If the runs comply exactly with Exception 5 to 310.15(B)(2), then they would not apply.

8. No. Ref.—396.12

9. No. It is allowed to be used where the support is no more than 6 feet (1.8 m) from the last point of support for connections within an accessible ceiling to luminaire(s) (lighting fixture[s]) or equipment.

10. Yes. 511.1 would cover the repair. Article 625 does not cover the vehicle, only the supply related to vehicle charging.

11. Yes. 342.30(B)(3), where threaded and threadless couplings are used.

12. Yes, if the threadless fittings are listed for the purpose. Ref.—344.42(A)

13. No. Ref.—322.10(3)

14. Yes.

15. Yes. Ref.—404.9(B)

16. Yes. Ref.—110.16

17. No. Ref.—334.12(A)(1)

18. No, however it may be a violation of the serving utility. Ref.—230.28

19. No. 514.11(A), (B) and (C), and 514.13 both are required.

20. No. Ref.—110.26(F)

2005 *CODE* CHANGES QUIZ 7

1. Yes. All branch-circuits that supply 125-volt, single-phase, 15- and 20-ampere outlets installed in dwelling unit bedrooms shall be protected by an arc-fault circuit interrupter listed to provide protection of the *entire branch-circuit.*

2. It would be considered as part of the lighting load. Ref.—220.52(A) Exception

3. It is not one of the permitted methods in 230.43 and is not permitted in wet locations. Ref.—320(3)

4. Yes. Ref.—250.104(B), last sentence.

5. Yes. Ref.—334.30(B)(2) except as prohibited by 334.12(A)(1)

6. No, only steel IMC is permitted by Article 342. Ref.—342.1

7. No, unless identified as sunlight resistant. Ref.—362.12(9)

8. Yes, all raceway articles permit cables unless specifically prohibited by the respective cable article. Ref.—342.22 through 362.22

9. Yes, 4 quarter bends (360 degrees total). Ref.—350.26

10. Yes. Ref.—300.50(B)

11. Listed and marked. Ref.—410.31

12. NUCC is nonmetallic underground conduit with conductors as covered by Article 354. Yes, it is required to be listed (Ref.—354.6). Yes, the conductors and raceway are required to be listed. (Ref.—354.100(B) and 354.100(C)).

13. No. Ref.—362.12(10)

14. In accordance with the installation instructions. Ref.—392.6(C)

15. The highest level that water can reach before it spills out. Ref.—680.2

16. EMT is a permitted equipment grounding conductor. Ref.—250.118(4)

17. These tables are provided to assist the user. The amount of temperature change would be provided by the user. Ref.—352.44 and the Tables 352.44 (A) and (B)

18. Yes. Ref.—378.6

19. Yes. Ref.—230.41(A)

20. 647.4(D). *Note:* FPNs are not enforceable there. This is the only location in the *Code* where voltage drop is mandated.

2005 *CODE* CHANGES QUIZ 8

1. No. Ref.—356.12(4)

2. 250.102, 356.60

3. No. Article 376.

4. 115 percent. Ref.—445.13

5. LCDI or AFCI. Ref.—440.65

6. No. Ref.—430.125(A) Exception permits alarm devices where the operation of the motor is vital to the safeguard of persons.

7. 30. Ref.—426.32

8. Yes. Ref.—404.9(B)

9. Yes. Ref.—404.8(A) Exception 2

10. No. It is identified for one conductor. Ref.—408.41

11. Intermediate metal conduit (Type IMC) and rigid metal conduit (Type RMC)

12. Yes. Ref.—424.3(A)

13. Battery or generator. Ref.—517.45(A)

14. Division 1. Ref.—500.5(C)(3)

15. 50 volts. Ref.—480.4

16. Purpose; use. Ref.—408.4

17. No. Ref.—430.245(B); 310.3

18. Thermal protection. Ref.—410.73(E)(4)

19. No. Ref.—440.14

20. Temperatures. Ref.—422.16(B)(3)

2005 *CODE* CHANGES QUIZ 9

1. No. Ref.—Article 100, definition of service; it must be supplied from the serving utility.

2. Article 110 Part IV. Ref.—Table of Contents

3. 6 feet (1.8 m); lower. Ref.—110.26(F)

4. 110.54(A) and 110.54(B). Ref.—Same

5. Three continuous white stripes. Ref.—200.6

6. Yes. Ref.—210.8(A)(2)

7. An arc-fault circuit interrupter *is a device intended to provide protection from the effects of arc faults by recognizing characteristics unique to arcing and by functioning to de-energize the circuit when an arc-fault is detected.* [Reprinted with permission from NFPA 70-2005.] Ref.—210.12(A)

8. No, that would be a feeder by definition. Ref.—Article 100—Definitions

9. False. Ref.—110.26(A)(3). Equipment located above or below other electrical equipment is permitted to extend up to 6 inches (150 mm) beyond the front of other electrical equipment.

10. 110. Ref.—110.26(F)j322

(1)(A)

11. 110, Part III. Ref.—Same

12. Three continuous white stripes on other than green insulation along its entire length. Ref.—200.6

13. GFCI (ground-fault circuit interrupter protection for personnel). Ref.—210.8(A)(2)

14. Arc-fault circuit interrupter (AFCI). Ref.—210.12 (B)

15. No. Ref.—210.52(B)(3). Some dwellings have more than one kitchen, and this new requirement was added to reduce overloading.

16. 3; 900. Ref.—210.52(D).

17. Two; readily accessible. Ref.—210.60(B)

18. 150 volt-amperes. Ref.—220.43(B)

19. 12; (3.7 m). Ref.—225.18(2); 225.18(2)

20. 100 amperes. Ref.—225.39(C)

2005 *CODE* CHANGES QUIZ 10

1. 230.40 Exception 4. Ref.—Same

2. 100 amperes. Ref.—230.79(C)

3. No, 240.4(D). Ref.—240.4(D)

4. 240.2. Ref.—240.2, definitions

5. Unlimited length. Ref.—240.21(B)(5)

6. 240.2. Ref.—240.2

7. They are required to be GFCI protected. Ref.—525.18(A)

8. First paragraph of 240.21, which states: *No conductor supplied under the provisions of Sec. 240.21 (A) through (G) shall supply another conductor except through an overcurrent device.* [Reprinted with permission from NFPA 70-2005.] Ref.—240.21, second sentence, first paragraph

9. No, the exception to 220.52 allows this load to be excluded from the calculation. Ref.—210.52 (B)(1) Exception 2 and 220.52(A) Exception

10. Yes, they would have to be accessible; however, only two would have to be readily accessible. Ref.—210-60(A) and (B)

11. No, they would still be accessible as defined by the *Code*. Ref.—Article 100, definition of accessible and readily accessible

12. Where the generator is a separately derived system. Ref.—250.20(D) and FPNs 1 and 2

13. There is no set distance other than it must be within 36 inches (900 mm) of the outside edge of each basin and it must be located on the wall immediately adjacent. Ref.—210.52(D)

14. Yes, an exception has been added to state that if the circuit supplies only a single bathroom, then it is permitted to supply other equipment within that single bathroom. Ref.—210.11(C)(3) and Exception

15. Yes, in places of assembly Article 518. Ref.—518.3(B) Exception

16. Listed. Ref.—695.10

17. No. Ref.—547.10(A)

18. No, an equipment grounding conductor in accordance with 250.118 is required, and the grounded (neutral) cannot be bonded to the grounding-electrode system where there are common metallic paths between the two buildings. Ref.—250.32(B) (1) and (2)

19. Yes, if it meets the conditions of 373.5(C) Exception. Ref.—Same

20. Yes, if they have a floor located at or below grade and are not intended as habitable, and are limited to storage areas, work areas, or similar use. Ref.—210.8(A)(2)

2005 *CODE* CHANGES QUIZ 11

1. Yes, or ground-fault protection is installed in accordance with 250.122(F)(2). Ref.—Same

2. A power panelboard is one having 10 percent or fewer of its overcurrent devices protecting lighting and appliance branch-circuits. Ref.—408.34(B)

3. Yes, in accordance with 110.14, 300.5(E), 300.13, and 300.15. Ref.—230.46

4. Yes, for feeders or branch-circuits. Ref.—590.4 (B) and (C)

5. Yes, in trade size ½ (16) through 1 (27) as a listed manufactured prewired assembly. Ref.—331.3(8)

6. Yes, only one is permitted generally. Additional feeders or branch-circuits are permitted in accordance with 225.32(A) through (E). Ref.—Same

7. Type AC cable with an insulated equipment grounding conductor sized to Table 250.122 is now allowed. Ref.—518.4(A)

8. Yes, they are not permitted to be installed in a face-up position in the countertops or work spaces. Ref.—406.4(E); 210.8(A)(7) and 210.52(C)(5)

9. The industrial portion of a facility meeting all the conditions of 240.2. Ref.—Same

10. Listed tamper resistant; listed tamper-resistant. Ref.—517.18(C)

11. Yes, if the luminaire (lighting fixture) does not exceed 6 pounds (3 kg). Ref.—314.27(A) Exception

12. At least 3 inches (75 mm) of conductor outside the opening. Ref.—300.14

13. No, the building codes may also prohibit combustible wiring methods. Ref.—334.12(A)(5)

14. It is not required for all buildings; industrial buildings are exempted. However, where the busway vertical riser penetrates two or more floors, a 4-inch (100 mm) minimum curb must be installed to retain liquids. Ref.—368.10(B)(2)

15. Two minimum where a made electrode is used and they have not been tested for compliance with 250.56. Ref.—250.52(A)(2)

16. No, only medium- and low-power systems are covered by Article 830. Ref.—830.4 and Table 830.4

17. Where a general-care patient bed location is served from two transfer switches on the emergency system, they are not required to have circuits from the normal system. Ref.—517.18(A) Exception 3

18. Yes, they could all be main power feeders if feeding lighting and branch-circuit panelboards. Ref.—310.15(B)(6)

19. Yes. Ref.—210.8(A)(3)

20. Article 314. Ref.—314.29

PRACTICE EXAM 1

1. Yes. Ref.—511.12

2. No. Ref.—680.72, however, 410.4(A) and (D) limit the use of pendant-type fixtures over this area.

3. Yes. Ref.—406.3(D)(3) and no, Ref. 406.3(D), but they must be marked "GFCI protected."

4. No. Ref.—210.8(A)(6)—only those serving the countertops. *Note:* Only very old ranges have receptacles mounted on them.

5. No. Ref.—314.3 Exceptions 1 and 2

6. Yes. Ref.—410.16(H)

7. 15-minute finish rating. Ref.—362.10 and 12

8. No. Ref.—300.4(B)(2)

9. Where run through the framing members, ENT shall be considered supported. Ref.—362.30(B)

10. Yes. Ref.—300.4(B) and 334.17

11. Yes. Ref.—334.112 and 340.10(4)

12. No. Ref.—517.10 Exceptions 1 through 3

13. Yes. Ref.—210.52(A)

14. No. Ref.—300.13(B)

15. Yes. Ref.—210.52(C) and 210.8(A)(6) *Note:* The GFCI requirement would not apply if the receptacle was located behind the refrigerator.

16. No. Ref.—210.52(A) and (E)

17. No. The requirement is now 6 feet 7 inches (2 m). Ref.—404.8(A)

18. See referenced exceptions and general rule. Ref.—410.31

19. General- and critical-care patient bed location. Ref.—517.18(B)

20. No. Ref.—210.52(A)—shall be in addition to those required if over 5½ feet (1.7 m) above the floor. See 210.8(A)(7)

21. Self-closing. Ref.—406.8(B)

22. No. Ref.—680.23(B)

23. Yes, provided it is installed and supported in accordance with Article 334. Ref.—334.6, 10, 12, and 30

24. No, they may be rubber or thermoplastic types. Ref.—225.4

25. Yes. Ref.—410.31

PRACTICE EXAM 2

1. No. Ref.—Only as environmental conditions require—547.1 and 547.8(A), (B), (C)

2. No. Ref.—320.30, 300.4

3. Extra-hard usage portable power cables listed for wet locations and sunlight resistance. Ref.—553.7(A) and (B)

4. Yes. Ref.—680.42(C) and 680.21(A)(1)—NM cable has a covered or bare equipment ground conductor, not insulated, but is now permitted with the dwelling occupancy.

5. Yes. Ref.—680.23(F)(1)

6. No, water heaters are not listed with an attachment cord. Ref.—400.7 and 8(1); 110.3(B)

7. No, generally. Ref.—312.8

8. Voltage with the highest locked rotor kVA per horsepower. Ref.—430.7(B)(3)

9. Yes. Ref.—680.25(A)

10. Yes. Ref.—210.21 and Tables 210.21(B)(2) and (B)(3)

11. Yes. Ref.—210.23(A)

12. Not enclosed, buried, or in raceway. Ref.—Table 310.17

13. No. Ref.—352.10 (FPN) is not mandatory, only explanatory

14. No, GFCI protection is still required. Ref.—590.6(A)

15. Yes. Ref.—210.8(A)(7)

16. No. Ref.—210.8(A)(7)

17. No maximum. Ref.—Table 220.14(J)

18. 6. Ref.—517.19(B)

19. Yes. Ref.—430.81(B) and 430.42(C)

20. Yes. Ref.—210.70(A) Exception 1

21. Yes, one or more supplied by a 20-ampere current. Ref.—210.52(F)

22. Yes. Ref.—Article 250 and Article 410 Part V

23. Yes, flexible metal conduit must be bonded to grounded conductor. Ref.—230.43(15)

24. No, only interior metal water piping systems. Ref.—250.104(A)

25. No, generally. Ref.—314.22, 314.16(A)

PRACTICE EXAM 3

1. No. Ref.—314.27(C)

2. No. Ref.—314.16(A) and (B)

3. No. Ref.—210.52(E) only single- and two-family dwellings

4. No, polarized or grounding type. Ref.—410.42(A)

5. Yes. Ref.—210.52(A)

Note: Due to the arrangement of furniture, the receptacle located behind the door may be the

only receptacle accessible for frequent use such as vacuuming.

6. No. Ref.—590.4(G)

7. 6, where integrated clamps, and the like, are not used. Ref.—Table 314.16(A)

8. Table 314.16(A) or 314.16(B) as applicable

9. Yes. Ref.—501.10(A)

10. No, however, it is a raceway. Ref.—Article 358

11. Yes, up to 5 feet (1.5 m). Ref.—358.30(A) Exception 1

12. Yes. Ref.—410.16(C)

13. No. Ref.—Definitions, Article 100, and 410.4

14. No, must be spaced 1½ inches (38 mm) from the surface. Ref.—410.76(B)

15. Yes, with exceptions. Ref.—410.65(C)

16. No, only where the grounded (neutral) conductor is present. Ref.—110.15, 230.56, and 408.3(E)

17. Yes, all kitchen countertop receptacles are required to be GFCI protected. Ref.—210.8(A)(6)

18. Generally no. Ref.—312.8 and Exceptions

19. Yes. Ref.—250.112(J); Article 410, Part V

20. No. Ref.—700.12(E)

21. No. Ref.—210.8(A)(2) Exceptions 1 and 2

22. Yes "all." Ref.—210.8(A)(1)

23. No; however, it cannot be counted as one of the countertop receptacles as required by 210.52(C). Ref.—210.52(B) Exception 3

24. No. Ref.—314.1 and 410.64

25. Within 12 inches (300 mm) of service head, goose neck, and at intervals not exceeding 30 inches (750 mm). Ref.—338.10, 230.51(A)

PRACTICE EXAM 4

1. 24 inches (600 mm). Ref.—Table 300.5

2. Yes. Ref.—Table 210.21(B)(2)

3. No, you are not required to locate required receptacles behind furniture. However, the number required by 210.52(A) must be installed convenient to furniture arrangement. At least two receptacles are required to be accessible. Ref.—210.60(B) and Exceptions

4. Yes, but only if the box is listed for fan support. Ref.—314.27(D) and 422.18

5. Ref.—410.31 and 410.32

6. Yes, generally. Ref.—410.65(C) Exceptions 1 and 2

7. Yes. Ref.—210.6(C)

8. Yes, it is required. Ref.—700.12(E)

9. 150 kVA or less. Ref.—Diagram 517.41(3)

10. Yes. Ref.—230.95

11. Yes, 15-volt luminaires (lighting fixtures). Ref.—680.43(B)(1).

12. Yes. Ref.—358.10(B)

13. No, it is not permitted. Ref.—250.58

14. No. Ref.—110.26

15. Yes. Yes. Ref.—210.8(A)

16. See 547.9 and 250.32.

17. No. Ref.—Part V, Article 230, specifically 230.70. *Note:* If service is properly established at the pole, the answer to this question is Yes.

18. 24 inches (600 mm). Ref.—Table 300.5

19. No, unless the cord and plug are part of listed appliance. Ref.—110.3(B), 400.8(1)

20. (a) No minimal burial depth requirements. Ref.—Article 810

    (b) As per 810.21 and 810.15

21. Yes, provided they control the power conductors in the conduit. Ref.—300.11(B)

22. Yes. Underground water piping system must be at least 10 feet (3 m) and be supplemented by an electrode type for it to qualify as an electrode. Ref.—250.52(A)(1)

23. Yes. Ref.—210.70

24. Entire fixture. Ref.—410.4

25. Yes. No, single conductors are okay; Type UF cable is not permitted. Ref.—225.6, 340.10, and 340.12(11)

PRACTICE EXAM 5

1. Yes. Ref.—392.3(E)

2. Ref.—410.8(D)

    (a) 12-inch (300 mm) incandescent, 6-inch (150 mm) fluorescent

    (b) 12-inch (300 mm) incandescent, 6-inch (150 mm) fluorescent

    (c) 6-inch (150 mm) incandescent, 6-inch (150 mm) fluorescent

3. No, generally four or less 14 AWG fixture wires are not required to be counted. Ref.—314.16(B)(1)

4. Yes, generally. Ref.—110.26(C)

5. 4 AWG copper. Ref.—250.66(B)

6. Yes. Ref.—700.12(E) Exception

7. Yes, if all conditions are met. Ref.—550.32(B)(1)–(7)

8. Yes. Ref.—680.25(B)

9. Yes, when smaller than ⅝ inch (15.87 mm), must be listed. Ref.—250.52(C)(2)

10. Yes, both are acceptable. Ref.—230.43

11. Exothermic welds. Ref.—250.70

12. No. Ref.—250.118

13. Yes. Ref.—518.4(A)

14. No. Ref.—430.21

15. Yes. Ref.—356.10 and 356.12

16. Yes, but bare conductors are permitted. Ref.—230.41 and Exception

17. No. Ref.—300.3(C) and 725.54

18. No, only circuits over 250 volts to ground. Ref.—250.97. See the Exception for listed boxes with a specified type of concentric knockouts that are acceptable.

19. Yes. Exceptions would permit this practice in some instances without raceway protection. Ref.—300.22(C)

20. Yes. Ref.—300.13(B)

21. 60 amperes. Ref.—430.72(B) Exception 2—Table 430.72(B) Column C

22. Yes. Ref.—406.10, 250.146

23. *Code* uses nominal voltages. Ref.—See definition, Article 100 or 220.5

24. 12 AWG THW conductor in cable or raceway. Ref.—Table 310.16 = 25 amperes

    Four conductors in a raceway Table 310.15(B)(2) —four to six conductors in a raceway or cable shall be derated to 80% = 25 × 80% = 20 amperes the allowable ampacity.

    Ambient temperature correction factors shown at the bottom of Table 310.16 require an additional deduction of .82 — 20 × .82 = 16.4 amperes. 210.21—outlet devices shall have an ampere rating not less than the load to be served. Table 210.21(B)(2) permits a 20-ampere branch-circuit for a maximum of 16 amperes. 16.4 exceeds this requirement. The conductor will carry the maximum allowable load. 240.4(D) limits the overcurrent protection on a 12 AWG conductor to 20 amperes.

25. 36" × 36" (900 mm × 900 mm). Ref.—324.10(H)

PRACTICE EXAM 6

1. Yes. Ref.—250.97, 250.94

2. Six. Ref.—230.71(A), 230.40 Exception 1

3. (a) Yes. Ref.—600.10(C)(2)

   (b) No. Ref.—600.10(C)(2)—There is no exception

4. Yes. Ref.—545.2, 210.70 (a dwelling), 210.70(C) (other types of buildings)

5. Yes, if they meet all the conditions. Ref.—605.8

6. Yes. Ref.—501.15(A), Article 500—Gasoline is a Group D, Class I hazard. Diesel is not. However, the sealing requirements are to prevent the passage of gases, vapors, and flames.

7. Yes. Ref.—210.52(A)

8. (a) 2 AWG copper or 1/0 AWG aluminum. Ref.—250.24 base on 300-kcmil Service-Drop Table 250.66

   (b) Bonded to the panelboard supplying each apartment size to Table 250.122 base and an overcurrent device 100 amperes 8 AWG copper or 6 AWG aluminum. Ref.—250.104

   (c) No. Ref.—250.52—A supplementary grounding electrode is only required where the underground water pipe is available. This installation is suggested by a nonmetallic water system.

9. Yes. Ref.—210.23(A)

10. No, generally. Yes, for interior portion of a one-family dwelling. Ref.—680.21(A)(4)

11. No, where an equipment grounding conductor and not over four fixture wires smaller than 14 AWG enter the box from the fixture. Ref.—314.16(A) Exception

12. Receptacle is not required specifically. 210.52(A) would require a receptacle near the wet bar. If located within 6 feet (1.8 m), a GFCI is required. Ref.—210.52(A), 210.8(A)(7)

13. Read heading for Table 220.55. Note 3 permits add nameplates of the two ranges. Use Column C. Ref.—Table 220.55 and Note 3

14. 4.5 kW. Ref.—220.14(B). 220.54 is used for feeder load calculations.

15. 35 pounds (16 kg). Ref.—422.18

16. Table 348.22 for trade size ⅜ (12). For ½ (21) through 4 (53) use Table 1, Chapter 9. Ref.—348.22

17. No. Ref.—314.23(B), 314.23(A)–(G).
    Note: 300.11(A) the general method requirements give limited permission for branch-circuits. However, junction boxes are covered by Article 314.

18. No, explosionproof not required or acceptable. Must be approved for Class II locations. Ref.—502.5

19. No. THHN is not acceptable in wet locations. Ref.—Table 310.13 and 310.8

20. Authority having jurisdiction. Ref.—90.4, also see definition in Article 100, "Approved"

21. Yes. Ref.—210.52(F)

22. Yes. Ref.—210.63

23. Column A is the basic rule to be applied when an exception does not apply. Ref.—430.72(B)

24. 83. Ref.—Duff and Herman, *Alternating Current Fundamentals, 4E*, Thomson Delmar Learning, and Chapter 3 of this book.

25. Qualified yes "Metal Multioutlet Assembly." Ref.—Article 380, 380.3

PRACTICE EXAM 7

1. Lighting circuit serving the area. Ref.—700.12(E)

2. Yes, if they are classified for purpose and meet all conditions of 392.7(B). Ref.—392.7(A) and (B)

3. No. Ref.—388.10(1)

4. If the required workspace is unobstructed or doubles. Ref.—110.26(C)(2)(a) and (b)

5. No, the anti-short bushing is not required. The conductors in MC cable are wrapped in a nylon wrapping. AC cable conductors are individually wrapped in kraft paper. Ref.—320.40, Armored Cable; 330.40, Type MC Cable

6. Generally, no, if exposed to physical damage. Ref.—362.10 and 362.12. *Note:* If these conditions do not exist, then ENT could be installed.

7. 18 inches (450 mm). Ref.—Table 300.5

8. Large capacity multibuilding industrial installations under single management. Ref.—225.32 Exception 1

9. No for Class 2 Div. 1, Yes for Class 2 Div. 2. Ref.—502.10(A) and (B)

10. No. Ref.—410.31

11. Grounded (neutral) conductor must be insulated generally. Ref.—310.12; 250.140 and 142. The rules for an insulated neutral or grounded conductor for a feeder, although not clearly written, the service neutral is often bare and 250.140 permits dryers and ranges originating in the service panel to be Type SE, which has a bare grounded conductor.

12. No. Classifying structures are the responsibility of the building inspector or the fire marshal generally. They rely on the adopted building code or the NFPA 101 Life Safety Code. Ref.—Article 518

13. No. Ref.—220.61, also Table 310.15(B)(4)

14. None. Ref.—Table 310.16, Ambient Temperature Correction factors at bottom of table.

15. Table 314.16(B)

Example: An outlet box contains a 120-volt, 20-ampere single receptacle and a 120-volt, 30 ampere receptacle. The 20-ampere receptacle is fed with a 12.2 W/GRD NM cable that extends from the outlet on to additional outlets located elsewhere. The 30-ampere receptacle is fed by an individual branch-circuit with 10.2 W/GRD NM cable. What size box is needed?

Answer: Ref.—314.16(A) and Table 314.16(B)

12.2 W/GRD feed	2.12 @ 2.25 cu. in.	4.5 cu. in.
12.2 extend through not spliced	2.12 @ 2.25 cu. in.	4.5 cu. in.
10.2 W/GRD feed	2.10 @ 2.5 cu. in.	5.0 cu. in.
30 ampere receptacle	2.10	5.0 cu. in.
20 ampere receptacle	2.12 @ 2.5 cu. in.	4.5 cu. in.
Ground	1 10 @ 2.5	2.5 cu. in.
Clamp	1 10 @ 2.5	2.5 cu. in.
		28.5 cu. in.

A box with 28.5 cubic inches is required.

16. 10 feet (3 m). Ref.—230.24(B)

17. No, it is considered by the *NEC®* as "Other Space Used for Environmental Air." Ref.—300.22(C)

and Exceptions, definition of "Plenum" in Article 100

18. No, this area must be unobstructed. Ref.—410.8 (A)

19. Equipment grounding conductor required. Ref.—250.134, 250.136

20. Yes. Ref.—410.4(A)

21. By interpolation. Ref.—430.6(A)(1)

22. 53 percent. Ref.—Chapter 9, Table 1 and Note 6

23. Yes. Ref.—Table 310.15(B)(6)

24. No, motors are computed in accordance with Article 430. Ref.—220.14, 680.3

25. Yes, if the garage is supplied with electricity. Ref.—210.8 (A)(2), 210.52(G)

PRACTICE EXAM 8

1. Yes, generally. Ref.—410.73(E)

2. As per 547.9 in compliance with Article 250. Ref.—250.32(E) (FPN) and 547.9

3. Yes, if known, and they occupy a dedicated space with the individual branch-circuit. Ref.—Table 220.55 "Heading." If they are plugging into one of the "small appliance" outlets, then they would not have to be added to the load calculations.

4. Yes, for controlled starting. Ref.—430.82(B)

5. No. Ref.—553.2 and 553.4

6. No. Ref.—Article 600, 210.6(A)

7. Yes. Ref.—210.21

8. Yes, limited application. Ref.—Articles 517.12, 13, and 30

9. No, generally. Ref.—240.3 and 310.15

10. Yes. Must be marked SWD. Ref.—240.83(D), 404.11

11. There shall be two deductions for the largest conductor connected to each device, and one deduction of the largest conductor entering the box for each of the following: the clamps, the hickeys, all of equipment grounding conductors, etc. Ref.—314.16(B)(1)

12. No. Ref.—250.94

13. The cable can be fished. However, 314.17(C) Exception requires stapling with 8 inches where a clamp is not used. Therefore, the answer is No. Ref.—334.30(B), 314.17(C) Exception

14. 150 amperes will meet all conditions required for a general-use switch. Ref.—440.12, 430.109(D)(3)

15. No. Ref.—348.60 and 348.20(A)(2)

16. I P/E   I = 10,000/240   I = 41.667
    R = E/I   R = 240/41.667   R = 5.76
    P = IE   P = 208 × 36.1   P = 7.51 kW
    I = E/R   I = 208/5.76   I = 36.1 amperes
    36.1 amperes—Chapter 3, Ohm's law
    I = P/E   I = 10,000/240   I = 41.667
    R = E/I   R = 240/41.667   R = 5.76
    I = E/R   I = 208/5.76   I = 36.1 amperes

17. Yes. Ref.—680.5, 680.22(A)(5), and 680.23(A)(3)

18. Chapter 9, Table 1 unless durably and legibly marked by manufacturer in cubic inches. Ref.—314.16(C)

19. By the markings on the luminaire (fixture) (IC) by the listing agency. Ref.—410.65(C) Exceptions 1 and 2

20. Conductors only. Ref.—240.21

21. Table 110.26(A)(1), Condition 3. Ref.—110.26 (A)(1)

22. Yes, if cord-and-plug-connected but not required or advisable. Generally, these appliances are wired to the requirements in Article 422. Ref.—210.52(B)(1) and (2)

23. 10 feet (3 m). Ref.—680.8(A)

24. Yes, no depth requirement but is required to be protected from severe physical damage. Ref.—250.64 and 250.120

25. Water meter and all insulating joints. Bonding around valves and sweated joints is not generally required but the water meter, filters, and insulating joints would be required to be bonded around. Ref.—250.52(A)(1) and 250.53(D)(1)

PRACTICE EXAM 9

1. Yes, but not advisable. One should isolate any possibility of imposing faults on those circuits. Ref.—680.5(B), 680.22(A)(5), 680.23(A)(3)

2. (a) Yes. Ref.—250.21 and 250.24(A) and (B)
   (b) In each of the parallel raceways. Ref.—250.24(A) and (B)
   (c) 2 AWG copper in each conduit. Ref.—250.24, 250.66

3. 21 inches × 24 inches × 6 inches. Ref.—314.28 (A)(2)

4. Yes, by Exception. Ref.—312.8

5. Yes. Ref.—358.30

6. Either system installation is okay where a three-pole transfer switch is used. Grounding electrode is not required for this installation. Ref.—250.24(D), 250.30

7. No. Ref.—300.5(I) Exception 2, although grouping of the conductors is preferred.

8. This installation is a violation; you cannot mix these conductors, unless the control is a Class 1 or power circuit. Ref.—300.3(C)(1), 725.55

9. Permit; however, it should not be located directly over the switchboard for the best coverage. Ref.—110.26(F)(1)

10. Yes. Ref.—250.50. Steel can be bonded to the water pipe within 5 feet (1.5 m) of where the water pipe enters the building. Ref.—250.50 and 250.52

11. 20-ampere breaker. Ref.—240.4(D) and 210.22

12. Yes. Also, the receptacles required by 210.52(E) still must be installed at direct grade access and be GFCI protected. Ref.—210.8(A)(3) and 210.52(E)

13. Yes, if listed for 2. Ref.—110.3(B) and 312.5(C)

14. No. Ref.—230.42(A), (B), and (C)

15. Yes. Ref.—430.109(F)

16. Article 310. Ref.—Table 310.15(B)(6)

17. No, neutral is not used at the switch. Ref.—200.7(C)(2) and 300.3(B)

18. Yes; but where they enter the building, they must be enclosed in a raceway. Ref.—340.10(4)

19. Yes, area may be unclassified by AHJ where there are four air changes per hour. Ref.—511.3(A) and 511.3(B)(1) Exception

20. No. Neutral must be capable of carrying the maximum unbalanced load. Ref.—250.140 and 220.22

21. Yes. Ref.—348.20(A)

22. Maybe; AHJ is the building inspector. Ref.—Local adopted building code—518.2(C)(FPN)

23. Yes, but it must be connected to the same circuit as supplies the lighting for the area to be served. Ref.—700.12(E)(4)

24. Maybe. This wet location is defined in 517.2; 210.8(B)

25. Yes. No. Ref.—422.12

PRACTICE EXAM 10

1. Yes, but storage of combustibles should be considered. Ref.—Part II, Article 410

2. Hot tubs and spas are defined in Article 680. Ref.—680.2

3. No. Ref.—511.4, 501.15(A)(4) Exception 2, 511.9

4. 150; No. P = EI = 480 × 104 = 49,920 single phase

   49,920 ÷ (480 × 1.73) = 60.12 amperes

   60.12 × 250% = 150

   Ref.—Table 450.3(B)(2)

5. Each section 5 feet (1.5 m) or fraction thereof shall be calculated at 180 VA in other than dwellings. Ref.—220.18(H)

6. No. Ref.—Article 430, Part IX, and 430.102(B)

7. Yes, possibly. Ref.—430.72 and Exceptions

8. Yes, except in an industrial facility. Ref.—368.11

9. No. Ref.—340.12(1) and 340.12(10)

10. Yes. Ref.—250.20(B)(2) and 250.24

11. No. Ref.—250.30 covers the requirements for a separately derived system. Nearest building steel is preferred.

12. No, you cannot tap a tap. A tap is defined in 240.4(E). Ref.—240.4(E), 240.21

13. Yes. Ref.—Part III, Article 424, 424.20(B)

14. 2 AWG. Ref.—Chapter 9, Table 8, Table 250.66

15. Yes. However, consideration must be made for corrosive elements. Ref.—300.6, 501.10(A), Article 344—Non-Ferrous Rigid Metal Conduit

16. No. Ref.—700.12(E)

17. No. Ref.—250.140

18. Yes, but there are installation requirements for disconnecting overcurrent. Ref.—230.82 and 695.3

19. Bond services together. Ref.—250.58

20. No. Ref.—680.71, 410.4(A)–(D)

21. No. Ref.—*NEC*®. Most requirements for GFCI protection appear in 210.8. Other requirements appear in Articles 511, 517, 527, 550, 551, 553, 555, and 680.

22. Yes. Ref.—410.28(D) and 410.31

23. Yes. Ref.—362.10(5). The 15-minute finish rating is a wall finish rating established by Underwriters Laboratories and can be found in the "UL Fire Resistant Directory."

24. Yes. Ref.—Article 314, 314.16(A) and (B), Table 314.6(B). 4 × 2⅛ = 30.3 cubic inches. 8 AWG conductor = 3 cubic inches. Therefore, ten 8 AWG conductors would be allowed without considering devices or other appurtanances such as clamps.

25. They must be fastened. Ref.—410.16(C)

PRACTICE EXAM 11

1. Yes. Ref.—340.10(4)

2. No. Ref.—680.23(A)(1)

3. Yes. Ref.—700.12(E) Exception

4. No. Ref.—300.22(C) Exception. Perpendicular, yes; parallel, no.

5. No, they must be equipped with a factory-installed GFCI. Ref.—600.11

6. In each disconnect means. Ref.—250.64(C)

7. No. Ref.—250.64(C) Exception

8. As a wireway in accordance with Article 376 or 378. Ref.—Article 376, Part II or Article 378, Part II.

9. Yes. Ref.—250.148

10. No. It can be insulated or covered. Ref.—680.21 (A)(4)

11. Yes. This installation is not in violation of the *Code*. Ref.—240.4(B), 310.16, 220.10, and 408.13.

    240.4(B)—The 400-ampere breaker properly protects 500-kcmil conductors that permit the next higher overcurrent device.

    220.10—Conductors 500 kcmil are large enough to carry the 180-ampere load.

    408.34—It is a power panel, not a lighting and appliance branch-circuit panelboard.

12. Yes, unless not required by NFPA 99. Ref.—517.25 (FPN)

13. Generally, no—Table 220.42, and 220.44. Ref.—220.12, 220.14, 220.40

14. Yes. No. Ref.—517.2 definition of patient care areas in hospitals. Ref.—517.13(A)

15. Yes, there are no conditions. Ref.—600.7, 250.112

16. No. Ref.—250.30

17. Yes. Ref.—310.4

18. They must be derated to 80 percent. Ref.—Table 310.15(B)(2)

19. Yes. Ref.—210.6(C)

20. Yes. Ref.—Definition of "Accessible Readily," definition of "Direct Grade Access"—210.52(E)

21. Yes, maybe. Ref.—210.52(D), 210.11(C)(3)

22. Yes, if they do not exceed allowable load of circuit. They are not continuous loads. Ref.—210.19(C) and 210.23

23. No. Ref.—210.52(G) only applies to dwellings (see definition, Article 100)

24. Yes. Yes. Ref.—210.70(A) Exception 2

25. Overcurrent device. Yes. Ref.—220.40

PRACTICE EXAM 12

1. Only branch-circuits. Ref.—220.51 and 424.3(D)

2. No. Table 220.55 Heading Column A applies to all notes as applicable, Columns B and C apply to Note 3. Ref.—Same

3. Yes. Ref.—225.30, 225.32, 225.33, 230.70, and 230.71

4. Location is not specified; where bonded to different electrodes, they must then be bonded together. Ref.—250.58

5. Yes, but may be wired as unclassified in accordance with Chapters 1–4. Ref.—514.1 Scope

6. ½ inch (13 mm). Ref.—410.66(A)

7. System ground. Ref.—810.21(F)

8. No. 90.3. Use Chapters 1–4. Ref.—511.3

9. Yes. Ref.—250.97 Exception

10. Yes. Ref.—680.21(A)(4)

11. No, only as taps (from raceway system to fixture). Ref.—348.20(A)(2)

12. No. Ref.—680.23(C)

13. No. Ref.—680.42, 680.43(C), and 680.43(D)(5)

14. No. Ref.—210.70(A)(2)(b)

15. 65 percent—Column C, Note 3—add 3 + 6 = 9 × 65% = 5.85 kW

16. Yes. Ref.—210.52(F). At least one but only for the laundry area.

17. Yes, all. Ref.—210.8(A)(1)

18. No, except when using Class 1 or power circuits for control. Ref.—725.55

19. Yes. Ref.—800.33

20. Bonding jumper shall be the same size as the grounding electrode conductor. Ref.—250.28 and 250.102

21. Yes, for wet bar sink; no, for laundry sink. Ref.—210.8(A)(7)

22. 25 foot tap = 1 AWG THW, 10 foot tap = 10 AWG THW. Ref.—240.21, Table 310.16, and 240.4(D)

23. No. Ref.—210.70(A). At least one switch

24. 75°C, because there are no circuit breakers rated 90°C. Ref.—110.14(C), 310.15(C)

25. Those areas determined by governing body of facility. Ref.—517.2 Definitions, Patient Care Areas of a Hospital

FINAL EXAMINATION

1. C. Ref.—800.14

2. B. Ref.—General knowledge

3. D. Ref.—General knowledge, *IEEE Dictionary*

4. A. Ref.—Chapter 3

5. A. Ref.—Chapter 3

6. C. Ref.—Chapter 3

7. C. Ref.—Chapter 3
   $P = I(E \times 1.73)$ $P = 83$ kW

8. A. Ref.—Chapter 3

9. A. Ref.—Chapter 3

10. C. Ref.—Chapter 3
    (16×4) [64] $\times (10 \times 4)$ [40] = 2560

11. A. Ref.—250.102(C)
    Note: 12½ minimum or not smaller than required in Table 250.66, individual bonding jumpers could be run from each conduit sized 1/0 AWG

12. A. Ref.—Chapter 9, Tables 1, 5, and 4
    2/0 AWG = .2265 sq. in. × 2 = .4530
    1 AWG = .1590
    .4530 + .1590 = .6120   Table 4 = 1½ inches

13. B. Ref.—Table 250.66

14. C. Ref.—230.95(A)

15. B with Exception. Ref.—230.9(A)

16. B. Ref.—Duff and Herman, *Alternating Current Fundamentals, 4E,* Thomson Delmar Learning

17. B. Ref.—Table 310.15(B)(2)

18. A. Ref.—Chapter 9, Note 3

19. B. 342.22 states for conductor fill see Chapter 9, Table 1. Ref.—Chapter 9, Tables 1, 4, and 5
    Solution: Table 1: over 2 conduits limits the raceway to a 40 percent fill.

Table 5:

4 AWG = .0824 sq. in. × 3 = .2472

1/0 AWG = .2223 sq. in. × 4 = .8892

Total = 1.1364 sq. in.

2″ IMC = 1.452   1½ IMC = 0.890

Therefore 2 (53) IMC would be required.

20. B. Ref.—352.22, Chapter 9, Table 1, Note 1 and Annex C, Table C10

21. C. Ref.—230.70(A)

22. C. Ref.—Appendix 1 (text)

23. D. Ref.—Appendix 1 (text)

24. B. Ref.—430.52

25. D. Ref.—Article 430, 430.37, 38, and 39, Table 430.37

26. B. Ref.—430.226

27. A. Ref.—Article 430, Part XI

28. D. Ref.—430.89

29. D. Ref.—250.140 (existing installations only)

30. B. Ref.—250.110(1)

31. C. Ref.—250.53(G)

32. D. Ref.—Article 430.7

33. D. Ref.—225.18

34. C. Ref.—230.95(C)

35. A. Ref.—430.24

Largest 5.6 × 1.25 = 7

4.5 × 1.00 = 4.5

4.5 × 1.00 = 4.5

16.0 amperes—minimum ampacity of conductors

36. B. Ref.—404.8(A)

37. D. Ref.—General knowledge; Annex 2

38. D. Ref.—General knowledge; Annex 2

39. A. Ref.—General knowledge; Chapter 3

40. B. Ref.—General knowledge; Chapter 3; Ohm's law; Annex 2

41. C. Ref.—Article 310

42. A. Ref.—Definitions, Article 100

43. D. Ref.—General knowledge; Ohm's law

44. B. Ref.—General knowledge

45. A. Ref.—Definitions, Article 100

46. D. Ref.—230.9(A) Exception

47. D. Ref.—Not a *Code* requirement

48. A. Ref.—Chapter 3

49. B. Ref.—General knowledge

50. D. Ref.—Ohm's law, 65 percent load, 220.56; Annex 2. Calculation is based on 1Ø units; if calculation based on 3Ø units, answer is "B".

Index